GLOBAL AGRICULTURE DEVELOPMENTS

VERMITECHNOLOGY

REBUILDING SUSTAINABLE RURAL LIVELIHOODS

GLOBAL AGRICULTURE DEVELOPMENTS

Additional books in this series can be found on Nova's website under the Series tab.

Additional e-books in this series can be found on Nova's website under the e-book tab.

GLOBAL AGRICULTURE DEVELOPMENTS

VERMITECHNOLOGY

REBUILDING SUSTAINABLE RURAL LIVELIHOODS

SHWETA YADAV
AND
V. K. SINGH

NOVINKA
New York

For permission to use material from this book please contact us:
Telephone 631-231-7269; Fax 631-231-8175
Web Site: http://www.novapublishers.com

NOTICE TO THE READER

Library of Congress Cataloging-in-Publication Data

Vermitechnology : rebuilding of sustainable rural livelihoods / editors: Shweta Yadav and Vinay Kumar Singh, Department of Zoology, School of Biological Sciences, Dr. H.S. Gour Central University, India.
pages cm. -- (Global agriculture developments)
Includes bibliographical references and index.
ISBN: 978-1-63117-943-3 (softcover)
1. Organic farming. 2. Vermicomposting. I. Yadav, Shweta. II. Series: Global agriculture developments.
S605.5.V47 2014
631.5'84--dc23
2014015861

Published by Nova Science Publishers, Inc. † New York

To
The Sacred Memory
of
My Father

Shweta Yadav

CONTENTS

FOREWORD

Agriculture with the use of agrochemicals in large amounts has, undoubtedly, resulted in a manifold increase in the production of farm commodities, but the adverse effects of these chemicals are clearly visible in the soil structure, soil microflora, quality of water, food, fodder and food materials. The World Health Organization has estimated that globally at least three million people are poisoned by pesticides every year, out of which 20,000 die. Traces of banned DDT and BHC isomers have been found in cereals, pulses, vegetables, human milk etc. Thus, organic farming is certainly an answer for creating a safe, good and clean environment, and the use of biofertilizers in preference to chemical fertilizers, offers economic and ecological benefits by way of soil health and fertility to farmers.

The compost prepared by using earthworms through vermitechnology is called *vermicompost*. Vermicompost is one of the organic nutrients such as Farm Yard Manure (FYM), Biofertilizers (both symbiotic and non-symbiotic), crop residues etc. Unlike chemical fertilizers, vermicompost is economic. It increases the productivity, water-holding capacity and nutritional value of soil and is absolutely harmless to the plants. Realizing the immense potential of vermicompost as a supplementary nutrient source of crops, a Vermitechnology was conceived in Aligarh (Western UP), India, way back in 2000 with a view to encouraging farmers of western U.P. to go for organic farming and provide themselves with 'state-of-the-art' technology for the production of vermicompost through setting up their own vermicomposting units on their own doorsteps in the villages.

It is heartening to note that the painstaking efforts of Dr. (Ms.) Shweta Yadav, Project Directress, have yielded tangible results. Over the years, the Vermiculture Research Station has developed into a fully-fledged R & D center, currently undertaking multifarious activities in different scientific pursuits with the support of the Government of India's UGC, DBT, DST & NMPB projects sanctioned during 2000-2012. Some of the major achievements of the center include the setting up of 1668 Vermicompost Units in 13 development blocks of Aligarh and Hathras districts with an annual production of 26.68 MT vermicompost. Demonstration plots on Good Agricultural and Collection Practices (GACP) for optimum doses of vermicompost and medicinal plants is yet another area in which the center is working hard, besides the promotion of cultivation and the marketing of high-demand medicinal crops. It is worth mentioning that the center has developed its distinct identity among the farmers of the region and has been providing them with recent updates on vermitechnology, agro-techniques for the cultivation of medicinal plants, information on schemes and financial assistance from the state/central government for their benefit.

The present book entitled,"Vermitechnology: Rebuilding Sustainable Rural Livelihoods" provides the recent updates on the vermitechnology in India and abroad and the achievements and activities of the Vermiculture Research Station, Aligarh, since inception to date in the context of Vermitechnology as the lab to land program in this region.

I am delighted to write the foreword to this publication and congratulate the authors Dr. (Ms.) Shweta Yadav, Associate Professor in Zoology, Dr H.S.Gour Central University, Sagar and Dr. V.K. Singh, Former Deputy Director (Botany), CCRUM, Dept. of AYUSH, Govt. of India, New Delhi, on producing this commendable work. Dr. (Ms.) Shweta happens to be a pioneer in spreading the concept of vermitechnology among the farmers of western U.P. through her expertise in the subject. This initiative has already started showing results by generating self-employment opportunities in the rural sector, particularly among the weaker sections of society, and is empowering women by improving their economic situation.

It is hoped that the information contained in this book will further alert and enlighten the farmers, scientists, professionals, entrepreneurs and others interested in vermitechnology as a tool for organic farming and looking to set up their own vermicompost production unit on a mass scale.

(Prof. N. S. Gajbhiye)
Vice Chancellor

Place: Sagar
Date:16.08.2013

PREFACE

Organic farming is a system approach utilizing the natural cycles and biological interactions for crop production and protection. It aims at achieving harmony with nature and production without destructing the environment. Vermitechnology is one of the important components of organic farming. India produces about 2,500 million tons of organic waste annually which could be utilized for the recovery of four essential commodities; namely, fertilizer, fodder, fuel and food. If properly managed, India could produce about 400 million tons of plant nutrients from this huge amount of organic waste. The earthworm (verm) is now known to be a good biological element for the recovery of vermifertilizer and vermin protein for use in agroecosystem, aquaculture, poultry etc. Earthworm culture, popularly called vermiculture, is being widely practised in a big commercialized manner. Therefore, it is high time for India to enter into the vermitechnology arena as part of the biotechnology program. In India, too, some companies have come up, but by and large general awareness remains limited, despite the fact that several non-governmental organizations and governmental institutions are trying hard to popularize vermitechnology for its adoption. This book is an attempt to help those who desire to start the vermiculture and vermicompost industry.

Based on this rationale, the Vermiculture Research Station, Aligarh, embarked on an extensive program of transferring vermitechnology to the farmers of the Aligarh and Hathras districts of western U.P. way back in 2000 under the Lab to Land program. The farmers were given specialized training through field demonstrations in vermitechnology, the organization of special courses and the setting-up of their own field units in rural areas. It is heartening to note that women farmers have also come forward in this new venture and are now producing quality vermicompost to supplement their income. This initiative has not only generated self-employment opportunities among the rural population of western U.P., but has also developed a sense of economic independence among the farming community. With our constant efforts, a series of vermicompost units have been set up with mass production and their assured marketing, through sale counters set-up, is now a reality. Four "Farmers' Exhibitions" organized in 2006, 2008, 2010 and 2011 provided the farmers with a recent update in vermitechnology and agro-techniques for the quality production of their produce. The center has been providing regular counseling to farmers since 2000, and has developed considerable awareness among them by working with them in the field and addressing their day-to-day problems.

It is hoped that the basic information on vermitechnology provided in this publication will encourage fellow scientists and farmers to take up work in this direction, which we hope would be in the best interests of the nation as organic farming is becoming an important alternative to present day high input intensive agriculture.

Our foremost thanks go to the farmers of western U.P. for showing great skill and efforts in learning the vermitechnology which has been imparted through the Vermiculture Research Station, Aligarh, since 2001. The encouragement and help provided by the farmers, the authorities concerned, different government funding agencies, such as UGC, DBT, DST, NMPB etc., and all the staff members engaged in the Vermiculture Research Station, Aligarh are gratefully acknowledged.

Dr. Shweta Yadav
Dr. V. K. Singh

About the Authors

Dr. (Ms.) Shweta Yadav: Associate Professor in Zoology, School of Biological Sciences, Dr. H.S.Gour Central University,Sagar (MP),India obtained her Masters in 1990 (Standing first in order of merit) from B.R. Ambedekar University, Agra (India) and Ph.D in 1994 on the thesis titled "*Microecological Studies of Soil Fauna*". Presently taking courses in Zoology at graduate & post-graduate level and guiding research for Ph. D. degree.

Recipient of 'Young Scientist Award' for consecutive years 1993, 1995, 2001, Dr. Shweta was adjudged for a prestigious 'Juagadishwari Rao Women Scientist Award (2002)'. She has been honored by Union Minister of Agriculture Hon'ble Shri Rajnath Singh in March 2003, for her outstanding contribution of setting-up series for Vermicompost Units in rural areas for production of organic manure, thereby generating self employment opportunities to weaker sections of the society.

Former member of the Task Force, Department of Biotechnology, Govt. of India, and associated with many learned scientific societies across the country, Dr. Shweta has to her credit over 58 research papers as a lead author, published in prestigious scientific journals in India and abroad. Over the years she has been invited to participate and make her presentations as key speaker in many national & international scientific meets including UK, USA, Potugese, Switzerland Malaysia and Nepal.

Dr. V. K. Singh: Former Deputy Director (Botany) at the Central Council for Research in Unani Medicine, Ministry of Health and Family Welfare, Government of India, New Delhi, is responsible for execution of projects n ethno-pharmacological field surveys, literature research, medicinal plants cultivation, development of herbal gardens and pharmacopoeial standardization of crude drugs. Dr Singh received his PhD in plant taxonomy. Based on his studies he has to his credit over 90 research & review papers published in reputed scientific Journals in India and abroad. Dr Singh is editor of various scientific journals and series editor for an international publication *Recent Progress in Medicinal Plants* (Vol.1--25), published by Sci. Tech. Publishing House, LLC, Texas, USA. Recognizing his contributions, Dr Singh earned several honors and awards including "*Lifetime Achievement Award, 2005*". Currently, Dr V.K.Singh is associated with an ambitious programme on awareness, training, field scale cultivation of medicinal plants including their organic farming in western UP districts in India.

Chapter 1

INTRODUCTION

The word 'Organic' means originating from a living thing. Organic farming is the production management philosophy of making the production system quality assured and free from all contaminants and pollutants. Organic farming is not just replacing fertilizers and pesticides with organic manure and biopesticides, but it is an organic dynamic process for making healthy soil and ultimately a vital living system for the world. Organic farming is the most widely recognized alternative farming system in the agricultural production context. It is a productive system, which avoids or largely excludes the use of synthetically compounded fertilizers, pesticides, growth regulators and livestock feed additives. To the maximum extent possible, the system relies upon crop rotations, crop residues, animal manures, legumes, green manures, off-farm organic wastes including soil humus, mineral-bearing rocks and aspects of biological pest control to maintain soil productivity and tilth, to supply plant nutrients and to control harmful insects, weeds, pathogens and other pests.

Despite spectacular progress in food grain production and the use of external chemical fertilizers the world over to meet increasing demand, the industry has to increasingly bear the escalating costs of raw materials, most of which are non-renewable in nature. At this stage when accelerating agricultural development requires further thrust, in many parts of the world it is becoming difficult to meet the nutrient needs of farming through fertilizer alone. Furthermore, the subsidy given on fertilizers in developing countries like India is becoming huge and unsustainable in the years to come (Table 1). It is in this context that organic farming is gaining ground day by day. The judicious and integrated use of a chemical source of nutrients along with nutrients through organic sources such as vermicompost, farm yard manures (FYM), crop residues, green manures, etc. will help reduce reliance on fertilizers and make good use of natural resources available locally to the farmers. The vermicompost is based on a renewable energy source and is cheap and environmentally friendly. Vermicompost can play a very significant role in improving soil fertility by enriching soil in microorganisms, adding plant hormones such as auxin, gibberellic acid and adding enzymes such as phosphatase and cellulase, attracting deep-burrowing earthworms already present in the soil; improving water-holding capacity, enhancing germination, better plant growth and higher crop yield.

Vermicompost (also called worm compost, vermicast, worm casting, worm humus or worm manure) is the end product of the breakdown of organic matter by some species of earthworms. Vermicompost is a nutrient-rich, natural fertilizer and soil conditioner.

Table 1. Subsidy given (10 million INR) by the Government of India on major fertilizers: 1990-1991 to 2008- 2009*

Year	Food	Fertilizers				Total Subsidies
		Indigenous Urea	*Imported Urea*	*Decontrolled fertilizers*	*Total*	
1990-91	2450	3730	659	-	4389	12158
1991-92	2850	3500	1300	-	5185	12253
1992-93	2800	4800	996	-	5796	11995
1993-94	5537	3800	762	-	4562	11605
1994-95	5100	4075	1166	528	5769	11854
1995-96	5377	4300	1935	500	6735	12666
1996-97	6066	4743	1163	1672	7578	15499
1997-98	7900	6600	722	2596	9918	18540
1998-99	9100	7473	333	3790	11596	23593
1999-00	9434	8670	74	4500	13244	24487
2000-01	12060	9480	1	4319	13800	26838
2001-02	17499	8044	47	4504	12595	31210
2002-03	24176	7790	-	3225	11015	43533
2003-04	25181	8521	-	3326	11847	44323
2004-05	25798	10243	494	5142	15879	45957
2005-06	23077	10653	1211	6596	18460	47522
2006-07	24014	12650	3274	10298	26222	57125
2007-08	31328	12950	6606	12934	32490	70926
2008-09	43627	16517	10981	48351	75849	129243

* Sources: Government of India (2009).

1.1. Need for Organic Farming

In the herbal sector, the produces from organic farming have become premium due to increasing market demand for standardized and quality products. Food, medicinal herbs and cosmetic products containing pesticides, heavy metals and other chemicals are increasingly reported nowadays. Many research studies have revealed the presence of pesticide residues which have horrified the consumers. About one million people suffer from pesticide poisoning and 20,000 die every year due to the toxic effects of chemicals worldwide. The value of crop loss due to pests is estimated to the tune of US$ 3-4 billion. In order to avoid the deleterious effects of synthetic chemical fertilizers, heavy metals like lead, cadmium etc., alphatoxins and pesticides, organic cultivation is imperative and the savior of the hour. It is an ecologically safe and environmentally sustainable method of farming.

1.2. Herbal Medicine

In recent years, there has been an overwhelming response to the organic produce in the world and it is growing at 20% per annum. In medicinal and herbal sectors this is absolutely necessary as all the major European, Japanese and North American buyers are demanding chemical-free raw materials and products. Therefore, the organic cultivation of medicinal plants is the current need. Hence, national and international efforts and research are needed to facilitate the successful adoption of organic cultivation and production of medicinal and aromatic plants. The approach has to be amenable to the farmer, collector, local traders and processors, with simple technologies at an affordable cost.

1.3. Global Market Trend

For the worldwide market for 'organically' grown, processed and certified food and herbal products, a coordinated national program and local actions are needed to promote the successful adoption of organic cultivation by poor and marginalized farmers. South Asia, with its rich agricultural diversity, stretched across 15 different agro-ecological zones and traditions of agricultural practices, has tremendous potential in practicing organic cultivation and production of medicinal plants, products and services. Most South Asian small and marginal farmers, especially in the hills and rain-fed areas, have traditional practices and are organic producers by default, as they cannot afford even the subsidized chemical inputs.

Wide varieties of organic products are sold worldwide (Figure 1) and the demand is growing rapidly in Europe, USA, Japan, and Australia. India is already exporting a range of organic products like tea, spices, rice etc. Furthermore, the organically grown products are commanding 40-50% kg/hectare price on average compared to conventionally given commodities.

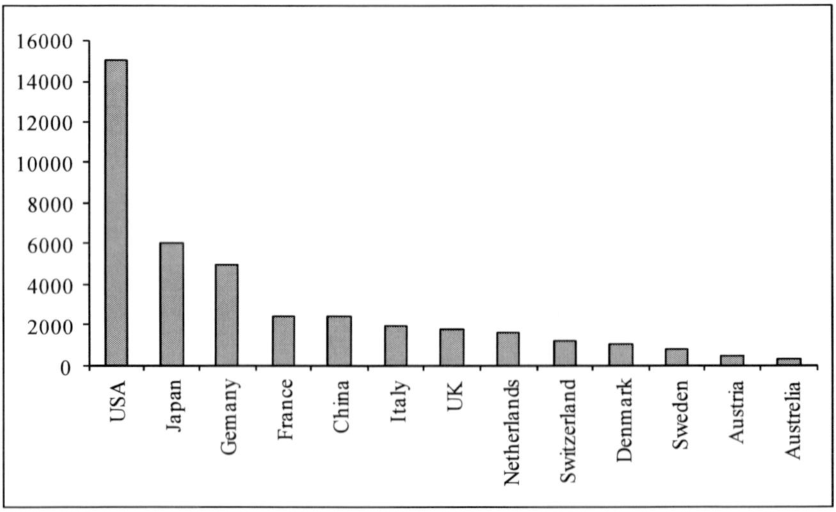

Figure 1. Value of organic food consumed in some countries ($ millions).

Even the Indian domestic market is growing rapidly, although it is still very small. In recent years, as hundreds of thousands of farmers[1] across the country (with an average land holding of four acres) switched over to organic farming, the organic production in India has considerably increased. Although one should not expect miracles from organic production, it is a realistic and promising option, especially for marginal and resource-poor farmers. For gaining consumers' confidence, valid certification is an essential prerequisite for marketing, especially in the export market. Organic inspection and certification are complex tasks, and regulations and accreditation criteria are manifold. Only a product recognized by nationally and internationally

[1] Source: *India Together*, Thursday, 03 April, 2008.

accredited certification agencies would get access to both export and domestic organic markets.

Internationally, the demand for organic medicinal, aromatic and dye yielding plants is growing with more and more people seeking health solutions from "alternative" and complementary medicine whose worldwide market is estimated at more than 60 billion US$ annually. Inclusion of cosmetics and pharmaceutical health products will perhaps push the value close to 310 billion US$. Greater awareness of the impact of quality production methods on the quality of a product will lead to a growing number of consumers choosing ecologically and particularly organically grown herbal products.

Chapter 2

SUSTAINABLE AGRICULTURE AND VERMITECHNOLOGY

The term '*Sustainable Development*' was coined by the Brundtland Commission Report '*Our Common Future*' in 1982 which redefined the concept of sustainable development. Sustainable development is the kind of development (both social & economic) that 'meets the 'needs' (but not the 'greed') of the present generation without compromising the abilities of the future generations. That should improve the total quality of all life (human beings, plants and animals) on earth, now and in the future too. While, maintaining the social and ecological integrity (natural and man-made ecosystems) of the earth upon which all life depends and which can provide good quality of life to all the people born on earth. The scientific community all over the world is desperately looking for an '*economically viable, socially safe and environmentally sustainable*' alternative to the destructive '*chemical agriculture*' which would not only 'maintain' but also 'enhance' farm production per hectare of available land as the farmlands all over the world are shrinking in the wake of rapid urbanization. Thus, it is not enough to produce 'sufficient food' to feed civilization (which was the primary objective of chemical-based green revolution) but also to produce a 'high quality of food' which should be 'safe' (chemical-free) and also 'protective' of human health (good combination of macro and micro nutrients and vitamins) and do it in a sustainable manner to ensure 'food security' for all.

Food Safety & Security is a major issue everywhere in the world. This will amount to embarking on a 'Second Green Revolution' and this time by 'Organic Farming' practices completely giving up the use of agrochemicals. The new concept of farm production against the destructive 'Chemical

Agriculture' has been termed 'Sustainable Agriculture'. This is about growing 'nutritive and protective foods' with the aid of biological based 'organic fertilizers' without recourse to agrochemicals. This is thought to be the answer to the 'food safety and security' for human society in future. The U.S. National Research Council defined sustainable agriculture as '*those alternative farming systems and technologies incorporating natural processes, reducing the use of inputs of off-farm sources, ensuring the long-term sustainability of current production levels and conserving soil, water, energy and farm biodiversity*'. It is a system of food production which avoids or largely excludes the use of systematically compounded chemical fertilizers and pesticides and uses environmentally friendly organic inputs. To the maximum extent feasible, organic farming systems rely upon crop rotations, crop residues, animal manures, legumes and green manures to maintain soil productivity and tilth to supply plant nutrients. It emphasizes both preventive and curative methods of pest control such as the use of pest resistant cultivars, bio-control agents and cultural methods of pest-control. In the US the top 25% of sustainable agriculture farmers practicing 'organic farming' now have better gross margins and better yields than the top 25% of their counterparts still practicing chemical agriculture.

Sustainable agriculture is synonymous with 'Cleaner Agriculture' as the objective is to reduce or even eliminate the use of dangerous agrochemicals from food production and also to reduce the use of other precious farm inputs like water and energy whose indiscriminate use to boost food production (to feed the growing masses) has led to widespread environmental destruction by way of soil salinity, waste and pollution. In India several farmers are being motivated to shift to 'organic farming and sustainable agriculture' through vermiculture and to give up 'chemical agriculture'.

A number of villages in India in the districts of Aligarh, Hathras, Agra, Meerut in Uttar Pradesh and Hazipur and Nalanda in Bihar have been designated as '*vermi-villages*' in which the farmers have completely embraced organic farming by using earthworms and vermicompost. They have completely given up the use of chemical fertilizers for the last ten years since 2003. They are growing cereals (rice, wheat and corn), fruits (guava, mango and lemons) and vegetable crops (potato, tomato, onion, brinjal, cucumber, and okra etc.) on vermicompost. Farmers of vermi-villages feel proud of their food products and they sell these at a higher price in the markets due to their good appearance and taste.

2.1. Historical Background of Agriculture in India

India, over several millennia, has been the treasure land of biological wealth, intellectual knowledge and spiritual wisdom. She has also been the cradle of many ancient human civilizations like the world-famous Indus valley, Ganges delta, Cauvery delta civilization etc. During both pre- and post-Vedic periods, arts, science and technology flourished well in various fields. Ancient mathematicians in India invented the 'Zero' and made remarkable contributions to the field of Algebra. An ancient astronomer named Aryabhatta discovered the earth's rotation and also the solar and lunar eclipses. Five planets were known to the Vedic Indians even before telescopes were invented.

Ancient physicians established two excellent medicinal systems; namely, 'Sidha' and 'Ayurveda', which are now recognized as 'healthy friendly' systems. A scientist named Bramagupta discovered a scientific theory synonymous with Newton's laws of gravity.

Similarly, the biological wealth of India could be evidenced in the report of Kothari (2001). India is classified among the 12 mega-diversity centers of the world in relation to crops. As many as 167 species of crops, 320 species of wild crop relatives and several species of domesticated animals have been identified. The genetic diversity within these species is outstanding. Some examples are rice (50,000 varieties), sorghum (5,000 varieties), mango (1000 varieties), pepper (500 varieties), cattle (27 breeds), sheep (40 breeds), poultry (18 breeds) and buffalo (8 breeds). This amazing diversity is not a fluke of nature, but a result of careful selection and even cross-breeding, over centuries, by India's farmers and pastoralists. Hence, Indian agriculture, to such an extent, is very traditional and dates back about 10,000 years. India, with its traditional crop varieties and crop production techniques, was able to feed its population and was exporting some selected produces to other continents. All the Indian villages were self-sufficient, self-sustained and self-reliant units. In fact, urban centers were dependent upon the villages, rather than vice versa.

When India attained independence, it faced a severe shortage in the food supply during the 1950s and 60s due to various reasons. To overcome this problem, India entered into the era of the 'green revolution'. Because of this, self-sufficiency in food production was somehow achieved with the introduction of so-called high-yielding varieties, inorganic fertilizers, pesticides, modern implements etc. These were imported from western countries in the name of 'technological packages' without taking account of their adaptability to Indian conditions and the possible consequences. The development of modern agriculture production was achieved by creating large-scale specialized farm production units, increased mechanization and the use of chemical inputs.

Apart from the substantial increase in the crop yield through the green revolution, these modern technologies have made the Indian farmers become poorer as they required a heavy investment in costly external resources with uncertain future availability. Furthermore, these technologies with the principle of monoculture resulted in the rapid erosion of crop and livestock diversity, natural soil fertility and biological pest regulation; they enhanced the soil erosion, salinization and environmental problems, and reduced the nutrition. Once obtained 'wild', foods on farms increased the need for expansive and poisonous chemical fertilizers and pesticides and finally made the farmers more dependent on markets and outside agencies. The impact was greater on tens of millions of small farmers and pastoralists.

Therefore, a section of policy makers and agricultural scientists now understand that a continuation of modern agriculture might lead to severe ecological and economic problems. They are also convinced that modern agriculture may not be able to meet the requirements of an ever-increasing population in the future. Hence, they are searching for alternative technologies. Several alternatives have been proposed such as low external input agriculture, organic farming, biodynamic farming etc.

In underdeveloped countries, the green revolution has resulted in boosting production due to intensive agriculture. Although it has resulted in good harvests and productivity by reaping three crops in a year with good irrigation facilities, there was no thought given to its adverse impact in the long run on the soil conditions and the environment, particularly during the last four decades. With the extensive use of chemical fertilizer and improper irrigation, the productivity of the soil is considerably reduced. The permanent and cheapest solution to overcome the dangerous effects of modernized agriculture is to develop a farming system which is economically productive and long-lasting in sustainable farming or natural farming by simple and inexpensive practices like vermibiotechnology. Thus, the old agricultural systems - namely, biological, organic, ecological, regenerative, natural, biodynamic and low input agriculture - are reconsidered for their sustainability.

2.2. Use of Fertilizers and Deterioration of Soil Environment

Fertilizers are combinations of chemicals which are used to increase soil fertility. The manufacture of these chemical fertilizers is highly energy intensive and uses a lot of fossil fuel as a source of energy. A crucial situation has arisen to minimize the use of fossil fuel in the manufacture of chemical fertilizers, as this is not an ever-lasting resource. It also involves the importing of ingredients for fertilizer production. Energy inputs to farms delivered by fertilizers range from 12 to 15 Mcal/kg N;0.6 to 1.5 Mcal/kg P_2O_5, and 0.5 Mcal/kg K_2O, and the average energy input for pesticides is 61,470 to 86,910 Kcal/kg active ingredient for insecticide, 62,770 to 99,190 Kcal/kg for herbicides and 27,770 to 64,910 Kcal/kg for fungicide. The runoff from the fertilizer factories causes severe atmospheric and water pollution. Precautionary measures have to be taken to avoid fire hazards and leakage of poisonous gases into the atmosphere from these units. This is an expensive proposition.

The primary ingredient of the phosphorus fertilizers is the deposit in the earth's crust. To minimize the cost of transport, the factories have to be set up near the mining sites. Only one-third of the raw material used in the production is recovered as fertilizer and the rest of the material is waste, which causes an acute disposal problem. Sufficient precautionary measures have to be taken to overcome the pollution hazards from these units. In general, the production of chemical fertilizers leads to the depletion of fossil fuel, causing environmental pollution and occupational hazards.

Fertilizer manufacturing has become a major activity of the chemical industry. It was thought that an increase in fertilizer production would be the first step to facing the challenges of food production. To avoid a shortage of food production it was considered necessary to use fertilizer to increase the yield. Now, along with the macronutrients, the application of micronutrients has to enter the scene since it has been realized that micronutrients are equally essential for the proper growth and yield of crops. A deficiency of any of these elements may affect the condition of the crops.

Though high levels of these chemicals are added to soil, the plants cannot absorb all that has been added to soil. They cannot remain in reserve for the next crop. These chemicals either get leached or form complexes with metal ions to form undesirable complexes. The constant use of chemicals increases their leaching because of the depletion in organic carbon. In the due course of time the accumulation of salt changes the pH of soils and they become unproductive. The crust formed in soil affects the porosity of the soil and becomes less productive due to the lack of good top soil.

To restore the soil's health, farmers are now inclined to revert to organic farming. With this changing trend it is necessary to produce organic manures in large quantities from the available sources of organic matter to maintain the high level of productivity from the agricultural land.

2.3. Feasibility of Organic Farming in India

Organic farming has the twin objectives of system sustainability and environmental sensitivity. In order to achieve these two objectives, it has developed some rules and standards which must be strictly adhered to. In India, with its ever-increasing population, there is increasing demand for food and with meager availability of organic resources, total organic farming is not possible. However, some specific areas can be diverted to organic farming for the export of high quality organic products, more specifically high value horticultural crops. Thus, on the whole under Indian conditions, only partial switching to organic farming is possible presently.

In Indian agriculture, the use of chemical fertilizers cannot be totally eliminated; rather, it can be reduced, or minimized. It has been proved by various experiments that by conjoint application of inorganic fertilizers along with various organic sources it is possible to sustain higher crop productivity, improved soil quality and soil productivity, besides meeting N, P and K requirements. These organic sources also help in alleviating the increasing incidence of deficiencies of secondary micronutrients. At present the commercial mineral fertilizers are the main sources of plant nutrients. Therefore, these organic resources should be used in integration with chemical fertilizers to narrow the gap between the addition and removal of nutrients by crop as well as sustaining the quality of soil and achieving higher crop production. The nationwide adoption of organic farming is not possible due to the unavailability of organic resources, reduction in productivity etc. which will leave many people hungry. In this context, renowned Agricultural Scientist and thinker Dr. M.S. Swaminathan (2003) said that "a hungry man is an angry man" and "*if the hungry man happens to be a young man, then we have a potential terrorist among us*".

Thus, in India, the adoption of organic farming is possible only for crops with a high export potential in international markets. On the other hand, full adoption of Integrated Green Revolution farming, another option to organic farming, is possible to a large extent, in areas with the basic features of the green revolution such as intensive use of external inputs, increased irrigation, use of high yielding and hybrid varieties, mechanizations, limited damage to the environment and human health. For this purpose some organic techniques are developed and combined with input technology to create integrated systems such as "Integrated Nutrient Management" (INM), "Integrated Pest Management" (IPM) and biological control methods which reduce the need for chemicals.

Hence, for developing countries like India, alternatives such as traditional methods have special advantages over modern agricultural techniques. Also, the capital and skill requirements in the use of traditional technologies are generally low and their adoption often requires very little restructure of the traditional societies. These traditional technologies are nothing but indigenous knowledge. In adopting such indigenous knowledge, our ancestors did not face any problems with large pest outbreaks or economic crises, unlike today's farmers.

Nowadays, natural farming, organic farming, eco-agriculture, permaculture, biodynamic agriculture, *Rishi Kheti, Sadhu Kheti* and so on generate a lot of curiosity among the general public. Although all these differ slightly in their precise meaning and emphasis, the underlying conceptual idea is the same. Their descriptions and methodologies are derived from a close study of nature. They all agree on the principle of least interference in natural systems to raise healthy plants and animals. They all agree to limiting and in some cases eliminating the use of artificial chemicals and heavy machinery in agriculture. They all aim at quality and the nutritious nature of the farm produce.

2.4. Vermitechnology: A Global Movement

The movement was started in the middle of the 20th century and the first serious experiment for the management of municipal/industrial organic wastes was established in Holland in 1970 and subsequently in England and Canada. Later vermiculture was followed in the USA, Italy, Philippines, Thailand, China, Korea, Japan, Brazil, France, Australia and Israel. However, farmers all over the world have been using worms for composting their farm wastes and improving farm soil fertility for a long time.

Abstract of Hon'ble Supreme Court of India Decision on Waste Disposal (Writ Petition No.888/1996)

Recommendations for Mandatory Action by Citizens

- Storage of waste at source
- Segregation of recyclable waste/non-biodegradable waste at source
- Conversion of organic waste into bioorganic fertilizer
- Collection, transportation and disposal of Hotel and Restaurant waste

Discretionary Recommendation

- ULBs should earmark a minimum fund on per capita basis for services e.g. 100 per capita per year

Suggestions for consideration of Central and State Government/UTs

- To set up a national technology mission for Solid Waste Management(SWM)

SWM to be on the National agenda

- Central and state governments to share funds with the local bodies and make them financially sound for implementation of the 74th amendment to the constitution

SWM infrastructure to be a pre-condition for all new urban development plans and proposals

- SWM infrastructure like waste collection and temporary storage sites for transfer processing and the disposal of urban solid wastes shall be made a pre-condition in urban planning and management plans by improvement trusts and urban department authorities etc. and shall be strictly enforced.

Composting to be deemed an agricultural activity

- So as not to be subject to the payment of non-agricultural conversion changes and requirements for converting agricultural to NA land for compost plants at land fill sites

Fiscal incentives and funds for compost plants

- Single-window fast track clearance for the setting up of compost plants and issue of various NOCs, permissions, finance required etc. for setting up plants.

- Social Infrastructure states for city waste conversion processes for priority finance and concessional rates of interest as well as for taxation and excise purposes.
- Compost/organic manure plants to be included in the definition of infrastructure facility and deemed a sanitation system. US 10 clause 23G(c)(ii) of the Income Tax act, similar to water supply and savage system.
- Excise exemption for plant and machinery intended for manufacture and testing of organic manure from urban solid waste and agricultural waste.
- Organic manure and vermicompost to be exempt from sales tax

The agricultural and fertilizer policy of the Central and State Govts/UTs should promote the use of city compost

- National and state Agriculture and Fertilizer policy should focus on revitalization of Indian soils by promoting the use of city composts within a reasonable distance of its production at a reasonable price acceptable to cultivators.

In the UK, large (1000 mt) vermicomposting plants have been erected in Wales. The American Earthworm Technology Company started a 'vermi-composting farm' in 1978-79 with 500 ton/month of vermicompost production. Hartenstein and Bisesi (1989) reported on the management of sewage sludge and effluents from intensively housed livestock by vermiculture in the USA. Japan imported 3000 mt of earthworms from the USA during the period 1985-87 for cellulose waste degradation. The Aoka Sangyo Co. Ltd. has three 1000 t/month plants processing waste from paper pulp and the food industry. This produces 400 tons of vermicompost and 10 ton of live earthworms per month. The Toyhira Seiden Kogyo Co. of Japan is using rice straw, municipal sludge, sawdust and paper waste for vermicomposting involving 20 plants which in total produce 2-3 thousand tons of vermicompost per month. In Italy, vermiculture is used to biodegrade municipal and paper mill sludge. Aerobic and anaerobic sludge are mixed and aerated for more than 15 days and in 5000 cum of sludge 5 kg of earthworms are added. In about 8 months the hazardous sludge is converted into nutritive vermicompost. In France, 20 tons of mixed household wastes are being vermicomposted everyday using 1000 to 2000 million red tiger worms (*Eisenia andrei*) in earthworm tanks. In Wilson, North Carolina, U.S., more than 5 tons of pig manure (excreta) are being vermicomposted every week. In New Zealand, Envirofert is a large vermicomposting company operating in a site of over 70 acres in Auckland converting thousands of tons of green organic waste every year into high quality compost (www.envirofert.co.nz).

Vermiculture is practiced and propagated on a large scale in India too as a part of the 'Agriculture Development Program' (to convert all the agriculture wastes into compost). Several organizations (including the government, private, NGOs etc.) have come forward to educate the farmers in the use of vermicast or *in situ* vermiculture. In this context, the Vermiculture Research Station, Aligarh (UP) India, standardized the procedure for the production of vermicompost and transferred the technology to farmers in different villages of Aligarh and Hathras districts through a series of educational training programs which has successfully transferred vermitechnology to farmers at grass-roots level. The Vermiculture Research Station, Aligarh, harbors a reasonable diversity of earthworm species such as *Eisenia fetida, Perionyx excavatus* and *Eudrilus eugeniae.*

2.5. Promotion of Vermitechnology for Sustainable Agriculture

Sustainable agriculture is a process of learning new and innovative methods developed by farmers and farm scientists and also learning from the traditional knowledge and practices of the farmers and implementing what was good in them and also relevant in present times. Vermiculture was practiced by traditional and ancient farmers with enormous benefits accruing for them and their farmlands. Therefore, there is a need to revive this 'traditional concept' through modern scientific knowledge - 'A Vermiculture Revolution'. Sir Charles Darwin called the earthworms 'farmer's friends'. There is great wisdom in this statement of the great visionary scientist who advocated the use of earthworms, 'nature's gift' in farm production. It is necessary to adopt and implement a food and agriculture production system which must ensure:

- Maintenance of soil microbiology and fertility by greater use of biofertilizers.
- High productivity and stability of yield over the years.
- Productivity with 'minimum' or 'no' tilling; 'low' use of agrochemicals and integration with biofertilizers and biopesticides.

- Productivity with minimum use of water and even sustaining dryness or heavy rainfall.
- Preservation of crop diversity (biotopes).
- Preservation of soil, water and air quality in the farm ecosystem.
- Preservation of benevolent organisms (predators) flora & fauna in the farm ecosystem.
- Preservation of groundwater table.
- Preservation of good health for all.
- Reduction of water and energy use.

These are the objectives of sustainable agriculture. Sustained vermiculture practices and the use of vermicompost in farm soil over the years would meet several of the above requirements for a truly sustainable agriculture. Vermicompost is rich in microbial diversity and plant-available nutrients; it would improve the moisture holding capacity of soils, thus reducing water for irrigation; improve physical, biological and chemical properties of soil; and increase soil porosity and softness, thus requiring minimum tillage.

Vermi-biotechnology is an ecofriendly, socially sound and economically viable innovative type of technology; to manage the organic waste resource on a low capital input basis does not call for expensive laboratories or sophisticated industrial instruments. It can provide employment to millions of young people; it can eliminate the dependence on chemicals, convert wastes into fertilizers, bring waste-land under cultivation, feed hungry citizens and make a country green and prosperous in the span of just a few years.

It is basically composting with earthworms (*Eisenia fetida, Eudrilus eugeniae, Perionyx excavatus)*; in nature all organic matters eventually decompose. It can be virtually anywhere; either indoor or outdoor conditions. Organic wastes used for composting are animal dung (cattle dung, sheep dung, horse dung, goat and poultry droppings), agricultural waste (after harvesting and threshing of the produce), forestry wastes (wood-sawing, peels sawdust and pulp), city leaf litter (mango, guava, oranges etc., from residential areas) and paper and cotton clothes (if these are not being recycled for useful products, they can be recycled with this technology) etc.

Usually 2 kg of earthworms recycle 1 kg of organic waste in 24 hours in ideal conditions. The worms recycle moist food of their own weight every 24 hours. Vermicomposting is simple and no special training is required. It can be started with very low start-up costs and can be done with the available infrastructure and space. The earthworms perform all processing naturally and the resulting product provides stable organic matter. It increases water-holding capacity and improves soil quality, pH, soil biological activities and the soil productivity in subsequent years.

2.6. VERMITECHNOLOGY AND SUSTAINABLE FOOD PRODUCTION

Earthworms have over 600 million years of experience in waste and land management, soil improvement and farm production. Sir Charles Darwin called them the '*unheralded soldiers of mankind and farmer's friend working day and night under the soil*'. Vermiculture biotechnology promises to provide cheaper solutions for:

- Management of municipal and industrial solid wastes (organics) by biodegradation and stabilization and converting them into nutritive organic fertilizer (vermicompost) as emphasized by Greek Philosopher Aristotle who called them the 'intestine of earth' which meant that they can digest a wide variety of materials from earth.
- Restoring and improving soil fertility and boosting food productivity by worm activity and use of vermicompost (miracle growth promoter) without recourse to the destructive agrochemicals.

Nations of the world today are seeking the most cost-effective, economically viable, environmentally sustainable and socially acceptable technology that can convert all organic waste into a valuable resource to be used back into human society. Earthworms have the potential of generating NPK equal to 10 million tons annually in India (and other nations too) as a huge amount of organic waste is generated every year and 1,000 tons of organic wastes can be degraded to 300 tons of nutritive vermicompost rich in NPK and all essential micronutrients by about a few million worms whose population almost doubles every year. The organic fraction of the MSW (about 70-80%), containing plenty of nitrogen (N), potash (K) and phosphorus (P), is a good source of macro and micronutrients for the soil. Vermicomposting of all waste organics, especially food and garden waste, from society and using the nutritive end-product to grow 'food' again will establish the concept of 'circular metabolism' for a sustainable society.

2.7. Environmental and Economic Benefits of Vermiculture

There will also be ample opportunity to reduce energy use and reduction of greenhouse gas (GHG) emissions in vermicompost production locally at farms by the farmers themselves. A huge amount of energy is used and GHG emitted at chemical fertilizer factories apart from the '*toxic and hazardous wastes*' that are generated. Farm energy requirements might be reduced by 40% by more efficient methods of food production through vermiculture technology. If there is a decline in the use of external inputs (agrochemicals), with more use of locally produced biofertilizer (vermicompost), the costs of food produced by farmers practicing sustainable agriculture will be reduced significantly. There will be more useful trees, more farm wildlife, increased groundwater in wells and ponds, cleaner non-polluted water bodies, more soft and nutritive soils with biological organisms in and around the farmlands in the farm ecosystem where sustainable agriculture is practiced by vermiculture. These will help boost the 'economic prosperity' of farmers.

2.8. Earthworms for Promoting Sustainable Agriculture

Earthworms are an important organism in the soil doing a great service for mankind for millions of years now. This combines immense social, economic and environmental values together, which is now being realized and recognized. A branch of biotechnology called '*Vermitechnology*' is emerging by the use of earthworms to solve various environmental problems from waste management to land (soil) improvement. Sir Charles Darwin, the great visionary biological scientist, highlighted its role in '*soil improvement and farm production*' a long time ago and the traditional farming community was also practicing vermiculture in their farms. Unfortunately, very little attention was given to it by post-Darwin biological scientists or modern agricultural scientists or even the farming community of the world, who saw 'agrochemicals' as a technological boon to produce more food in a shorter time.

Biological and agricultural scientists all over the world, after being utterly disappointed by modern chemical agriculture which is destroying the soil and also adversely affecting human health (the 'boon' turning into 'bane'), are now looking back to the 'traditional wisdom' and trying to revive the dreams of Charles Darwin.

Earthworms, when present in soil, inevitably work as 'soil conditioner' to improve its physical, chemical and biological properties and also its nutritive value for healthy plant growth. This they do by soil fragmentation and aeration, breakdown of organic matter in soil and release of nutrients, secretion of plant growth hormones, proliferation of nitrogen-fixing bacteria, increasing biological resistance in crop plants and all these worm activities contribute to improved crop productivity. Worms swallow a large amount of soil with organics everyday and digest them by enzymes. Only 5-10 per cent of the digested material is absorbed into the body and the rest is excreted out in the form of fine, mucus-coated granular aggregates called 'vermicastings' which are rich in NPK (Nitrates, Phosphates and Potash), micronutrients and beneficial soil microbes. The value of earthworms in plant propagation was emphasized by the great Indian author Surpala in his epic '*Vriksha-ayurveda*' (Science of Tree Growing) as early as the 10th century A.D. He recommended incorporating earthworms in the soil of pomegranate plants to obtain high quality fruits. This traditional wisdom has been scientifically verified today for the successful and sustainable growth of several fruits, vegetables and cereal crops without the use of agrochemicals.

2.9. Vermicompost – Quality Manure

Vermicompost is a stable and fine granular organic matter. When added to a clay soil it loosens the soil and provides the passage for the entry of air. The mucus associated with the cast being hygroscopic absorbs water and improves water-holding capacity. Sandy soils, which have less water retention capacity, are greatly benefited by vermicompost application. The strong mucus-coated aggregates of vermicompost help soils to hold water for a longer time. The organic carbon in vermicompost releases the nutrients slowly and steadily into the system and enables the plant to absorb these nutrients. The soil enriched with vermicompost provides additional substances that are not found in the chemicals. In a nutshell, vermicompost improves the physical, chemical and biological properties of soil in the long run on repeated application. The properties of vermicompost which make it an ideal biofertilizer for the soil are:

i. Vermicastings have immobilized enzymes like protease, lipase, amylase, cellulase, lichenase and chitinase which help in the biodegradation of macromolecules of the agricultural residues in the soil so that further microbial attack is accelerated.

ii. Vermicompost is rich in vitamins, antibiotics and growth hormones.
iii. Vermicastings are free from pathogens.
iv. Vermicastings have immobilized microflora which help the soil to produce useful products.
v. Vermicastings present in vermicompost have earthworm cocoons which promote the earthworm population in the soil, thus ensuring the continuous production of vermicastings in the soil itself.
vi. It gives structural stability to the soil.
vii. It absorbs moisture from the air and holds it in the soil.
viii. Some of the secretions of worms and the associated microbes act as a growth promoter along with other nutrients. The multifarious effects of vermicompost influence the growth and yield of crops.
ix. Management of the live population in the field first necessitates the development of suitable conditions which are not easy due to modern agriculture practices. Even if these are made available, large cultures of suitable earthworm species (approximately 1 million live individuals per hectare land) are required. All such requirements are necessary for the development of earthworm cultivation (Vermiculture).
x. Huge quantities of domestic, agricultural and rural industrial organic wastes can be recycled for various stages. This also reduces pollution.
xi. Vermicastings are effective biofertilizers containing beneficial soil microflora and earthworm cocoons. When covered with a layer of mulch in the soil, they produce an earthworm population of one lakh per acre in 3 months. Activation of soil biology in this way is an essential component of natural farming.

2.10. Some Facts about Vermitechnology

i. The art of composting has been part of our global culture since ancient times.
ii. Worms live where there is food, moisture, oxygen and a favorable temperature.
iii. In one acre of land, there can be more than a million earthworms.

iv. Worms tunnel deeply in the soil and bring subsoil closer to the surface mixing it with the topsoil. Slime, a secretion of earthworms, contains nitrogen. Nitrogen is an important nutrient for plants. The sticky slime helps to hold clusters of soil particles together, called aggregates.
v. Worms are cold-blooded animals.
vi. Baby worms are not born. They hatch from cocoons smaller than a grain of rice.
vii. Even though worms have no eyes, they can sense light, especially at their anterior (front end). They move away from light and will become paralyzed if exposed to light for too long (approximately one hour).
viii. If a worm's skin dries out, it will die.
ix. Worms are hermaphrodites. Each worm has both male and female organs. Worms mate by joining their clitella (swollen area near the head of a mature worm) and exchanging sperm. Then each worm forms an egg capsule in its clitellum.
x. Vermicomposting is the easiest way to recycle food wastes and is ideal for people who do not have an outdoor compost pile.
xi. Composting with worms avoids the needless disposal of vegetative food wastes and enjoys the benefits of high quality compost. It is done with "redworms" (*Eisenia fetida*) who are happiest at temperatures between 50° and 70°F and can be kept indoors at home, school, or the office.
xii. Worms process food quickly and transform food wastes into nutrient-rich castings.

Chapter 3

Present World Scenario of Vermitechnology

The green revolution of India during the 1960s is well-known throughout the world. It helped to feed the hungry masses. India attained self-sufficiency in food production by adopting hi-tech agriculture in terms of high-yielding varieties, high fertilizer and pesticide dose, etc. About half of the rice, wheat, maize and vegetable areas in third world countries has been planted with modern varieties and fertilizer consumption in those areas has increased rapidly. Nitrogen consumption in the form of fertilizers has increased from 2 to 75 million tons in the last 55 years. The intensive use of fertilizer has brought the uncultivated land into cultivation. The fertilizer consumption which was 0.52 kg/ha in 1950-51 increased to 91 kg/ha during 2010-2011. As expected, fertilizer nutrient demand went up from 0.29 million tons in 1960-61 to 13.9 million tons at the end of 1995-96 (Chakraborty et al., 2002). During 2010-2013 the production and distribution of biofertilizers in the country was roughly 10,000 metric tons.

3.1. Production of Vermicompost in India

The use of *Perionyx excavatus* was the beginning of a pioneering attempt in India to employ earthworms for the degradation of solid waste. Later *Eudrilus eugeniae* was brought to the laboratory in the University of Agricultural Sciences, Bangalore, to test the possibility of their establishment in this subcontinent which has similar climatic conditions to those of its place of origin. Since 1982, *Eudrilus eugeniae* has been promoted for waste degradation. When compared to *Perionyx excavatus,* this species has surpassed it both in feeding and reproductive rate. In the beginning, it was thought that the response may be short-lived, as many organisms show a tendency to perform extremely well in a new environment until they get established. Such a change in their efficiency has not been observed in the many years since. A series of experiments has shown that it is an ideal species to work on agricultural and agro-based industry waste. Their sensitivity to density pressure and meandering nature demand a larger area of space for their establishment. At the same time *Eisenia fetida* also entered into Indian laboratories from Germany and is the preferred species for composting activity in urban areas (urban waste).

The *University of Agricultural Sciences, Bangalore*, India initiated vermiculture in India and propagated the knowledge of vermiculture to the farming community. The university released the technology to the farmers in 1984. At the initial stage, several controversies were developed about the composting and selection of species of earthworms. Most of the researchers were of the opinion that indigenous species of earthworms should be used for composting instead of exotic species. The group that opposed the use of exotic species in this country lacked the knowledge of their feeding habits, habitat preference, reproductive potential and their ability to survive in captivity to accept them for vermicomposting. But a detailed and in-depth study of the biology of 3 out of 14 species of earthworms collected during the survey of middle Gangetic plains of India helped us to put forward our views that a very few species of earthworms can be used for composting purposes and a few more as co-inhabitants to hasten the process of decomposting along with other soil organisms. In India, which was under colonization for several centuries, exotic species of plants and animals have been transported either intentionally or unintentionally. The northern part of the country has more exotic species of earthworms than the southern peninsula.

Agriculture Colleges under Jawaharlal Nehru Krishi Vishvavidayalaya, Jabalpur, Kasturbagram and *Indore* also propagated the vermiculture technology to farming community in the state of Madhya Pradesh. There are many other organizations which have been active in popularizing vermicompost in many parts of the country.

The *Green Cross Society of Mumbai*, a non-profit organization, has set up several projects to convert organics into vermicompost. A smaller-scale unit with a capacity of 4 tons of slaughterhouse waste per day, and a large unit with a capacity of 20 tons of vegetable waste per day, has been set up in Bombay, in collaboration with the Bombay Municipal Corporation.

Indian Aluminum Co. Ltd., Belgaum, Karnataka has initiated the processing of garbage and sewage from 500 homes. Compostable garbage is processed in seven concrete vermicomposting bins (one for each day of the week). Sewage is fed to a 200 square meter "vermifilter", a bio-filter with a 30 cubic meter depth. The vermifilter can process up to 100 cubic meters of sewage per day, and the purified water is used to irrigate gardens. This low-labor system needs little operation or maintenance.

M.R. Morarka – GDC Rural Research Foundation, Jaipur is popularizing vermicomposting in Rajasthan. They have conducted around 300 demonstrations on vermicompost in about 150 villages on various crops like bajra, wheat, barley, gram, moth bean etc.

Vermiculture Research Station, Aligarh is constantly striving to work in western UP with farmers and other groups to develop profitable vermicomposting projects or businesses. We worked with farmers to develop practical solutions for organic resource management with a view to finding the right balance of materials and methods to meet the goals.

3.2. Production of Vermicompost in Other Countries

World-wide scientists and farmers are establishing large-scale vermicomposting units. The Grace McKellar Centre, Geelong, Victoria, Australia vermicomposted about 13 cubic yards per week of kitchen scraps, shredded paper, and garden trimmings and projected savings are $14,000 per year. Hobart City Council, Tasmania digested 66 cubic yards of municipal biosolids per week, with the help of earthworms. They also mixed zeolite with the feedstock to balance the pH and to absorb ammonia and odor. About two-thirds of this volume becomes vermicompost, which is sold to the public on site. They plan to expand this operation to process 330 cubic yards per week. Newcastle City Council, New South Wales, Australia is currently saving $56,000 per year from the vermicomposting of avoided landfill tipping fees, and they are also receiving an equal amount of revenue from their sales of vermicompost. The resulting vermicompost is harvested weekly, bagged and sold to major supermarkets in Melbourne and Sydney.

National Institute of Environmental Health Sciences (NIEHS), Research Triangle Park, North Carolina, United States purchased two heated worm composting units; namely, *Worm Wigwams* and 10 pounds of worm in 1996. Shredded confidential office paper and finely shredded wood rejected for animal bedding for NIEHS's labs is used for bedding. Staff volunteers take turns bringing 10 to 20 pounds of food scraps each day from the cafeteria to the worm bins, located outdoors. Employees are using the worm castings as a soil conditioner for their plants.

Oregon Soil Corporation, Beaverton, Oregon, United States uses an automated vermicomposting system to process 12 to 14 tons per day of yard trimming and pre-consumer food scraps collected from supermarkets at a charge of $65 per ton (landfill tipping fees are $80 per ton). The organic materials are ground up and then mechanically applied in three-inch layers by a retrofitted manure spreader on a gantry which travels over the bed of 120 x 8 x 2.5 foot (960 square feet) in a *continuous flow reactor*. Vermicompost is produced in this unit in 30 days only and sold for $25 per ton. About 3,000 tons per year of the vermicompost is sold as "Oregon Soil All-Purpose Planting Mix" in one-cubic foot bags in a chain store in four or five states. This pilot-scale reactor costs about $40,000 to construct; another $50,000 was spent to build a greenhouse-like structure and buy a leachate tank and other equipment.

Pacific Southwest Farms (PSF), Ontario, California, United States is operating a large-scale vermicomposting unit. PSF has 360 windrows measuring 8 feet X 100 feet on their 54-acre site. More than 100 tons of worms are fed four tons of foodstock per week (100 tons per day) of municipal solid "green" waste brought from materials recovery facilities (MRFs). This includes tea, untreated wood, food-contaminated paper, and natural fiber products. About 45 percent of their income is from sales of casting to agricultural users.

In 1994 *San Quentin Prison California, USA* added vermicomposting to its Recycling and Salvage Program (RASP). Starting with two worm bins, within three years they expanded to 15 bins, all constructed from salvaged scrap wood from the prison recycling center. In 1996-97, worms consumed 400 pounds of food scraps weekly, with a production of 11,000 pounds of castings and 17,000 worms. Food scraps from the prison ranch kitchen are collected and buried daily in the worm bins, and papers collected from all offices at the prison are shredded and added to the bins as bedding. One or two inmates are assigned the work of the vermicomposting center each year to maintain and harvest the bins. Every four to six months, the bins are harvested for casting and excess worms. Although most of the castings are used by the prison landscaper in San Quentin's garden areas and flower beds, leftover castings are placed into buckets recovered from the prison's main mess hall and sold to prison staff to use on their home plants and gardens. Prison staff may also purchase harvested worms to use as fishing bait or to start their own worm bins. Prison officials are pleased with the success of the vermicomposting program, as it not only diverts food scraps from the landfill and provides the prison with a valuable soil amendment; it also produces an income that supports their recycling program.

Table 1. Expansion of organic farming area (ha) in Europe

Countries	1999	2000	% share of hectares	Number of farms
Italy	958687	1040377	7.01	51120
Germany	452279	546023	3.20	12732
UK	240000	527323	2.85	3563
Spain	352164	380838	1.11	13424
France	316000	370000	1.31	9260
Austria	272635	271950	8.42	19031
Sweden	155674	171682	3.70	3329
Denmark	146685	165258	5.5	3466
Finland	137000	147423	6.79	5225
Portugal	47974	50002	0.18	763
Ireland	32478	32355	0.69	1014
Netherlands	22997	27820	1.48	1391
Greece	15849	24800	0.64	5270
Belgium	18572	20263	0.94	628
Luxembourg	1002	1030	0.81	51
Total	3,169,996	3,777,144	2.94	130,267

More than five tons per week of swine manure solids are being vermicomposted at a farm *in Wilson, North Carolina*. Manure that has passed through an automated solids separator between the swine house and lagoon is placed on a 15 x 15 foot concrete pad. A front-end loader transports the manure to a 30 x 200 foot greenhouse, and a spreader delivers manure to 2 x 190 foot wooden worm beds which hold thousands of pounds of red worm. Temperature and moisture are controlled through the use of greenhouse curtains, shade cloth, fans, and an automatic mister. Castings are lifted and conveyed from the beds every other month by a retrofitted machine, and a harvester separates worms and eggs from the castings. Castings are sold in bags marked "Vermicycle: Nature's Ultimate Plant Food". Less than eight weekly person-hours are required to manage this operation.

Resource Conversion Corporation/Canyon Recycling, San Diego, California, USA is producing 35,000 tons of vermicompost from the yard trimmings of San Diego Zoo and Wild Animal Park and construction and demolition debris. Their vermicompost, which is marketed under the trade name "VenniGro", sells for $35 per cubic yard (regular compost sells for $5 per cubic yard).

Rideau Regional Hospital, Perth, Ontario, Canada holding a 725-bed hospital, with 1000 staff members, decided to implement vermitechnology using 825 pounds of wet organic produced each day. They constructed 20 bins, each 3' x 4' x 30", filled them with shredded newspaper and soil bedding, and added a total of 100 pounds of red worms. The beds are located in a greenhouse, which has to contend with temperatures ranging from 86°F degrees to 22°F.

Since 1991, *Sovadec, La Voulte, France* has been operating vermicomposting of 20 tons of mixed household waste per day. A thermal process opens plastic bags filled with refuse, and the contents are separated by automated equipment and hand sorters. Textiles, paper, cardboard, plastics, glass and metals are recycled, and the remaining materials are composted aerobically to eliminate pathogens, weeds and other harmful organisms. Next, the organic materials are processed in '*lombricubateurs*' (earthworm tanks) which each have a capacity of 15 tons per day, using 1,000 to 2,000 million red worms of *Eisenia andrei*. The resulting vermicompost meets most stringent European standards; non-organic material is less than 50 mm in size.

Vermiculture Production Center, Pinar del Rio Province, Cuba uses cow manure as its primary feedstock, in addition to pig and sheep manure, sugar cane pulp, coffee pulp, and other crop residues. The feedstock is thermophilically composted in twice-weekly turned windrows for 15 to 30 days before being fed to the worms. Vermicomposting takes place in windrows covering four hectares of orchard land. Thin layers of feedstock are spread by a tractor-drawn manure spreader. When the windrow reaches a height of 65 cm, worms are drawn to the surface with a layer of fresh feed, and then five to seven days later a front-end loader is used to skim the top 10 cm off the windrow, removing 80 per cent of the worms. The worm harvest can be brought to 90 to 92 per cent with a second feeding and loader pass. A mechanical harvest is used to separate the worms from their castings. Castings are used to replace manure as a fertilizer for tobacco; they are also applied to corn, tomatoes, garlic, onions and coffee bush nursery stock.

3.3. Some Success Stories of Indian Vermi-Entrepreneurs

In India, the green revolution has resulted in boosting production due to intensive agriculture. Although it has resulted in good harvests and productivity by reaping three crops in a year with good irrigation facilities, there was no thought given to its adverse impact in the long run on the soil conditions and the environment, particularly during the last four decades. With the extensive use of chemical fertilizers and improper irrigation, the productivity of the soil is considerably reduced.

Vermicomposting activity is being implemented successfully in the villages of the Aligarh and Hathras district of Uttar Pradesh in India with the assistance of the Department of Biotechnology, Government of India. A regular farmers' meet was organized to create awareness of the production and application of vermiculture as well as the establishment of entrepreneurship among the rural people. The project was implemented with the technical assistance of Dharm Samaj P.G. College, Aligarh (UP) in five Panchayat blocks; namely, Lodha, Jawa, Khair, Tappal and Dhanipur of District Aligarh and Hassain, Mursan of District Hathras of Uttar Pradesh. In Rajpur village of Tappal Block, a progressive farmer, Shri Asha Singh, is producing the vermicompost in bulk and personally applying upto 50 60 mt per month in the fields of paddy and vegetable crops. More than a hundred farmers of the Aligarh district have applied vermicompost in their fields and are not only saving upto fifty per cent of chemical fertilizers but also receiving 10-15% enhanced yield. According to the farmers' self-assessment of their farming, vermicompost is 4-5 times more effective than the conventional compost and this has developed competition among the farmers to produce greater and greater crop yield through vermicomposting (Biotech News Letter, 2003). A progressive farmer of Khushal Gari village, Atrauli Block, Aligarh owns a nursery of horticulture plants and has earned an income of around US$ 2500 through the production and sales of earthworms and vermicompost besides meeting his own nursery requirements. He also experimented with the use of vermicompost in the grafting of horticulture plants and observed a major change in accelerating the rooting of grafted plants.

Some enterprising graduate students have established their own vermiculture units, which enable them to meet their contingency requirements. Realizing the success of this program, rural banks of district authority have come forward to finance and associate themselves with the rural community to work for the technology dissemination and to encourage more and more farmers to adopt vermicomposting. The establishment of hatcheries and the availability of suitable earthworm quantities are the key factors for awareness building and the popularization of vermicomposting in rural areas.

Mass rearing and maintaining worm cultures and the tapping of organic wastes for their maintenance has a good scope for developing it as a cottage industry in developing countries like India to mitigate the problem of affecting crop diversity due to the intensification and use of inorganic fertilizers and pesticides. The tapping of this resourceful technology is of utmost importance in the present day "as soil is the placenta of life" to meet the requirement of rural India. This approach would help in developing a biovillage cluster. The country has no dearth of organic wastes, congenial climatic conditions and required manpower.

The self-help group approach is developing in a major way in vermicomposting in all the blocks of the Aligarh district and village banking institutions are highly impressed with the micro-credit system established by the village community. Good linkages have been developed in the village cluster for marking the worms and compost produced through cooperative building and self-help groups. Trained beneficiaries are being financed for the establishment of vermicompost units and vermiculture to propagate the activity in the village on a large scale.

3.4. Developing Women Vermi-Entrepreneurs

In order to develop women's vermicompost projects different villages' farmers and respective "Gram Sabhas" of the villages - namely, Devinagla, Alinagar, Balukhera, Jalupur Syore, Faridpur, Kheda Narainpur, Pankhani, Bhadasi, Gyanpur etc. of the Aligarh district of Uttar Pradesh - were surveyed. Pankhani (Dhanipur block), Bhadasi (Lodha block), Balukhera (Dhanipur block), and Gyanpur (Chandosi block) were selected for the purpose. As such, some 15 women (family members) of Pankhani, 17 women (family members) of Balukhera, 3 women of Bhadasi and 4 women of Gyanpur (family members) from these villages agreed to take up the vermicompost projects in their own houses. The women were provided with 8 kgs earthworms for two beds, and tools (one Hazara and one punja). Women farmers of these villages are using vermicompost in their own rabi and kharif crops.

The painstaking efforts made in developing women vermi-entrepreneur units during the study period have brought tangible results by way of: (i) setting-up a hatchery-cum-demonstration unit, comprising 60 vermibeds; (ii) providing 4000 kg vermiseeds to beneficiaries; (iii) organizing 13 training programs on vermitechnology in which some 600 women farmers participated; (iv) the setting up of 270 vermicompost units by women farmers in rural areas; (v) production and sales of 19 mt vermicompost through sale-purchase-counters by SHG of women farmers, fetching an earning of Rs. 5 lakhs annually; (vi) development of two 'vermi-villages' for the sustainable supply of earthworm seeds to new upcoming beneficiaries; (vii) organizing an exhibition, bio-fairs, and meetings about the project of vermitechnology, providing recent updates by experts, in which some 4350 farmers and others participated; (viii) participation of project in the Govt. exhibitions, "Krashak gosthi", meetings to educate, train and provide information on vermitechnology through interaction with farmers and exhibits, models etc; (ix) analysis of field soil where vermicompost has been used; and (x) new initiatives on the organic cultivation of vegetable and horticultural crops by SHG of women farmers, fetching high prices.

Although the objectives of the vermi-project have been achieved far and wide, and participating women farmers of the study area (Aligarh) feel privileged and empowered with vermitechnology at their fingertips, yet there is still a necessity to continue these efforts in adjacent districts to run this program in order to ensure the participation of women in this new venture.

3.5. Analysis of Soil Constituents of Farmers' Fields

Vermicompost has been produced by the Vermiculture Research Station, Aligarh UP over the years. As one of the project objectives, we have made an attempt to analyze some 100 soil samples collected from women farmers' fields in which vermicompost was applied. The study has revealed that there is an increase in carbon percentage from 3.28 to 6.19 in these fields. Furthermore, the pH of the same fields was decreased from 9.88 to 8.69 and an increase in nitrogen (196 mg/100g to 252 mg/100g), potash (13.2 to 21.6 mg/100g) and phosphorous (89.4 mg/100g to 219.3 mg/100g) was observed.

This indicated considerable improvement in soil constituents with the application of vermicompost.

3.6. Honoring Farmers for their Exemplary Works

Six farmers were honored for their exemplary works in the area of vermitechnology and medicinal plants cultivation at 'Bio-Fair-2008' at the Vermiculture Research Station, D.S. College, Aligarh, UP, India.

1. *Shri Asha Singh*, a progressive farmer from village Rajpur, Khair, Aligarh UP has earlier been the recipient of the first prize from the U.P. government in 1997 for sugar cane production. Shri Singh has taken an exemplary interest in vermicomposting since 2002 and has set up over 50 beds thus far, producing over 600 Quintal vermicompost per month. He has marketed the produce at Rs. 250 per Quintal earning Rs. 7500.00, besides using the vermicompost in his own crops. Shri Singh also multiplied the earthworms and sold over 60 kg @ Rs. 100/- per kg. Thus, he has been successful in generating self-employment through the encouragement, support and technology provided by the Vermiculture Research Station, D.S. College, Aligarh and has encouraged fellow farmers.
2. *Shri Ikram Ali,* an ambitious farmer from Hassain Vikas Khand, Hathras, Uttar Pradesh, took up organic cultivation of the medicinal plant Ashwagandha, applying vermicompost during 2007. He has produced 55 kg roots and sold these at Rs. 50/- per kg, earning Rs. 2750/-. Shri Ali observed that there has been no effect of frost on the crop produced through vermicomposting, whereas other crops of mustard and potato cultivated with the application of chemical fertilizers were severely affected by frost. Shri Ikram Ali has been a source of inspiration to other farmers who are now initiating the organic cultivation of medicinal crops.
3. *Smt. Kamlesh Devi* has been a very hard-working, progressive farmer. She received 38 kg earthworms from Vermiculture Research Station, Aligarh UP during 2007 and produced 33 Quintal vermicompost/ month. She is selling 8 Quintal vermicompost monthly at Rs. 200/- per Quintal and using the other 25 Quintal for her potato crop. Her

efforts have yielded good results and she has become a role model for other women of the village.

4. *Smt. Shkuntla Devi*, who has great enthusiasm, is from Dhanipur village in the Aligarh district of UP. She received 2 days' training in vermitechnology at Vermiculture Research Station, Aligarh, during 2007. In the first phase she prepared 2 beds and has been able to obtain 4 Quintal vermicompost. She encouraged 37 fellow women of Chhitupur village who have now taken up production of vermicompost on a mass scale.
5. *Smt. Rajni Devi*, a woman from the BPL group, has been working since 2007; she had been given 132 kg earthworms through Vermiculture Research Station, Aligarh. She prepared 33 beds, giving a production of 152.50 Quintal vermicompost/month and selling at Rs. 200/- per Quintal. She used fifty percent of their production in her own crops of potato and wheat. Smt. Rajni Devi has become very popular and has encouraged other women of the village to start producing vermicompost on their own doorsteps in the villages.
6. *Shri Dhyan Pal Singh and Shri Nathi* Singh are dedicated farmers and have been continuing their efforts to develop expertise in medicinal plants cultivation and marketing since 2002.

As a follow-up to the achievements of the National Medicinal Plants Board project (2002 – 2006) "*Awareness, training and cultivation of medicinal plants in western U.P.*", the farmers have formed the "*Herbal Agro Development Gramoudyog Society*". This is a commendable step to ensure the marketing of medicinal produce and vermicompost. The society includes 40 members of the farming community. The society has taken up plantation of 25000 plants of Amla in a 250 hectare area and cultivation of turmeric, garlic, jatropha, ashwagandha, kalmegh, gheekwar and more in 8 hectares of land. A 'mandi' is also being constructed for sale, along with herbal produce, vermicompost, and horticulture produce. These efforts of the community through Shri Dhyan Pal Singh and Shri Nathi Singh, who have encouraged fellow farmers in the cultivation of medicinal plants, deserve appreciation.

Chapter 4

VERMITECHNOLOGY: A SIMPLE TECHNIQUE

Vermitechnology involves the stabilization of compostable organics under controlled conditions by particular earthworm species. Compostable organic materials such as agricultural wastes, municipal wastes, non-toxic solid and liquid industrial waste and household garbage can also be converted into vermicompost. It is rich in nutrients and can be used as a valuable soil conditioner. A large number of enterprises in the commercial and industrial sectors have the potential to utilize vermiculture for the on-site treatment of compostable organics. Vermiculture technology requires careful planning and implementation to ensure the processing of organic materials with minimum impact on the surrounding environment.

4.1. ORGANIC MANAGEMENT SYSTEM

An organic management system involves the processing of a compostable organic material into an end product. Organic processing technologies include vermiculture and composting to produce vermicast or vermicompost. Establishing an organic management system for vermiculture processing involves much more than simply applying for composting. The selection of organic material, the development of a recipe for mixing the material, and harvesting of completed vermicast, management and monitoring procedures for the vermiculture units require careful planning to ensure the successful and efficient establishment of a vermiunit.

4.2. Basic Science of Vermiculture

Vermiculture is the processing of compostable organic materials in an aerobic (high oxygen) condition performed by worms and microorganisms. When organic materials are loaded into the vermiculture bed, bacteria, fungi and other microorganisms start to decompose or 'eat' them. The worm works in association with these microorganisms. Worms scavenge and eat the decomposing products such as sugars, proteins and simple carbohydrates released by the microorganisms. A well-managed vermiculture unit involves the maintenance of conditions that are ideal for the survival and growth of microorganisms and worms. Ideal environmental conditions include:

(i) Air: at least 10% oxygen (O_2) should be present in the vermibed for actively feeding the worms (there is 21% O_2 in normal air).
(ii) Moisture: moisture content of the bedding material should be between 60 and 90%. Ideal bedding temperatures are between 20 and 25°C; however, worms can survive between 5° and 35°C.
(iii) Food: such as discarded fruits, vegetables, wasted mixed food organics, paper, cardboard, industrial and animal wastes may be used for vermicomposting.

4.3. Components of a Vermiculture Unit

A vermiculture unit consists of four main components:

(i) Container: It houses the bedding and worm population so that the worms can consume organic materials loaded into the unit. The container excludes pests and predators.
(ii) Worm Population: Worms are necessary to convert the compostable organic materials into vermicompost.
(iii) Bedding: This is a deep layer of organic materials used as a medium for worms to live in. A greater mass of bedding reduces the variation in bedding temperature and moisture.
(iv) Mulching: This is the material applied on the surface of the bedding to meet the food and moisture requirements of growing worms.

4.4. Types of Vermiculture Units

A number of different types of vermiculture units are in use for the on-site processing of compostable organic material. These include:

4.4.1. Small-Scale Units

Small-scale systems use a wide variety of bins. Often, small-scale composters build their own bins. Commonly, bins are made of old plastic containers, wood, Styrofoam containers, or metal containers. Metal containers often conduct heat too readily, are prone to rusting, and may release heavy metals into the compost.

Bins should have holes in the sides to allow air to flow, and a spout that can be opened or closed or holes in the bottom to drain into a collection tray. Plastic bins require more drainage than wooden ones because they are non-absorbent. The design of a small bin usually depends on where an individual wishes to store the bin and the way they wish to feed the worms. Most small bins can be grouped into one of three categories:

(a) Non-continuous – *an undivided container.* A layer of bedding materials is placed in this bin as lining at the bottom. Worms are added and organic matter for composting is added in a layer above the bedding. Another layer is added on top of the organic matter and the worms will start to compost the organic matter and bedding. This type of bin is very common as it is small and easy to build. But it is relatively difficult to harvest because all the materials and worms must be emptied out while harvesting.

(b) Continuous Vertical Flow – a series of trays stacked vertically. The bottom-most tray is filled first, in a similar fashion to any other bin, but it is not harvested when it is full. Instead, a thick layer of bedding is added on top and the tray above is used for adding organic material. Worms finish composting the materials in the bottom tray and then migrate to the one above. When a sufficient number of worms have migrated, the vermicompost in the bottom tray can be collected and should be relatively free of worms. These bins provide an easier method of harvesting.

(c) Continuous Horizontal Flow – a series of trays lined horizontally. This method also relies on the earthworms migrating towards a food source and eases the process of harvesting. The bin is usually constructed to be similar to a non-continuous bin but is longer horizontally. It is divided in the middle, usually by a large gauze screen of wire. One half is used until it becomes full, then the other half is filled with bedding and organic matter. In time, the worms migrate to the side with the food and the compost can be collected. These bins are larger than a non-continuous system but still small enough to be used indoors, with the added bonus of being easier to harvest.

4.4.2. Large-Scale Units

There are two main methods of large-scale vermiculture – Windrows and Flow through system.

(a) Windrows System – Windrows systems are a traditional, low technology method for large-scale vermiculture activities (Fig.1a). They consist of long beds placed directly on the ground with compostable organic material being applied to the surface and sometimes covered to reduce the incidence of pests. Windrows systems are relatively inefficient as nutrients are not lost through volatilization and leaching and they require large areas of land.

It consists of bedding material for the earthworms to live in and acts as a large bin; organic material is added to it. Although the windrow has no physical barriers to prevent worms from escaping, they generally do not escape due to the abundance of organic matter on which they feed. Often windrows are used on a concrete surface to prevent predators from killing the worm population. When large-scale windrows are fed on one side consistently, a wave motion is generated over time.

A thatched roof shed, preferably open from all sides with an unpaved non-concrete floor, is erected in an east-west direction lengthwise to protect the site from direct sunlight. A shed area of 12’x12’ is sufficient to accommodate three vermibeds of 10’x3’ each having 1’ space in between for the treatment of 9-12 quintals of waste in a cycle of 40-45 days. The length of the shed can be increased or decreased depending upon the quantity of waste to be treated and the availability of space. The height of the thatched roof can be kept at 4 feet from the center and 2 feet from the side. The base of the site is raised at least 6 inches above the ground to protect it from flooding during the rains. The vermibeds are laid over the raised ground. The site marked for vermibeds on the raised ground is watered and a 4"-6" layer of any slowly biodegradable agricultural residue (such as dried leaves/straw/sugarcane or trash etc.) is placed over it after soaking with water. This is followed by a 1" layer of vermicompost or farmyard manure. Earthworms are released on each vermibed at the following rates: (i) 1.0 kg per vermibed for the treatment of cowdung /agri-waste; (ii) 1.5 kg per per vermibed for the treatment of household garbage.

The frequency and limits of loading the waste can vary depending upon the convenience of the user, as given below:

Frequency -Daily -2" /bed/day
Loading - In bulk -12-15" (3-4q/bed/cycle of 45 days)

The loaded waste is finally covered with a jute mat to protect the earthworms from birds and insects. Water is sprinkled on the vermibeds daily according to requirements and season to keep them moist. The waste is turned upside down fortnightly without disturbing the basal layer (vermibed). The appearance of black granular crumbly powder on top of the vermibeds indicates the harvest stage of the compost. Watering is stopped for at least 5 days at this stage. The earthworms go down and the compost is collected from the top without disturbing the lower layers (vermibed). The first lot of vermicompost is ready for harvesting after 2-2 ½ months and the subsequent lots can be harvested after every 6 weeks of loading. The vermibed is then loaded for the next treatment cycle.

(b) Flow Through System or Raised Bed System

Here the worms are fed an inch of "worm chow" across the top of the bed, and an inch of castings are harvested from below by pulling a breaker bar across the large mesh screen which forms the base of the bed (Fig.1b). As red worms are surface dwellers and are constantly moving towards the new food source, the flow-through system eliminates the need to separate worms from the castings before packaging. Flow-through systems are well suited to indoor facilities, making them the preferred choice for operations in colder climates.

4.5. Harvesting the Compost and Worms

There are three basic ways to separate the worms from the finished compost. One way involves moving the finished compost and worms over to one side of the bin and adding new bedding material and food waste to the other side. Worms in the finished compost should move over to the new bedding with the fresh food waste. The finished compost is then removed.

Figure 1. (a) Windrows System (b) Flow Through System.

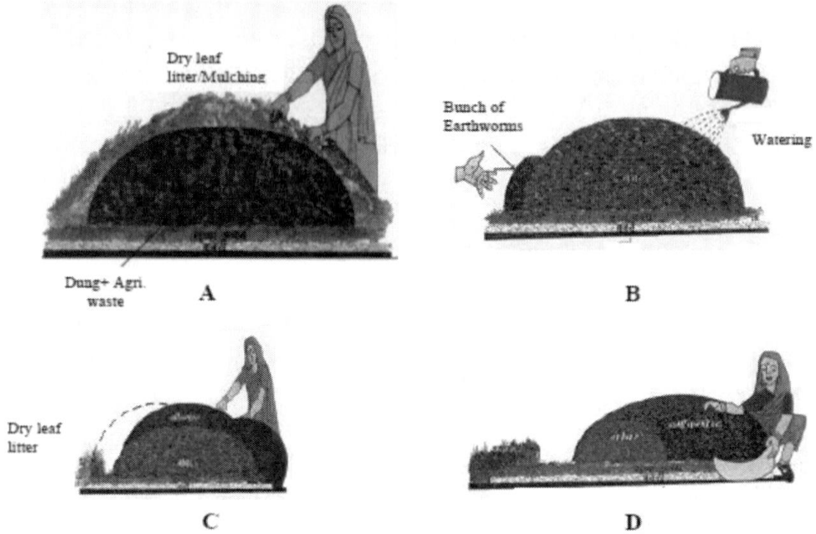

Figure 2. Steps of Vermitechnology, A: Bedding, B: Watering, C: Harvesting, D: Collection.

A second way to remove the worms is to build a small harvester frame and place the worm compost on the frame and shift the worms out. Larger pieces of compost can be returned to a new batch of bedding and worms (Figure 2). The compost can also be placed in small piles on a trap in the sun (or under bright lights inside). Because worms don't like light, they will wiggle to the bottoms of the piles. After waiting 10 minutes, remove the upper inch or more of finished compost from each pile until you run into the worms. Allow the worms to again wriggle to the bottom of the pile and repeat the process. Combine what's left of the small piles into one big pile and again repeat the process. This should eventually result in a pile of finished compost and a ball of worms. The worms can be added back to a new bin of bedding and food waste. Larger worms can also be used as bait for fishing.

4.6. Materials That Can Be Processed to Produce Vermicompost

Vermiculture units can be used to process an unlimited range of compostable organic materials into a useful end product called *vermicast*. However, effective vermiculture processing requires significant management of the unit to ensure reliable performance and to prevent health and environmental issues. Effective management of the vermiculture unit requires a dedicated approach for feedstock preparation, monitoring regimes and site hygiene. A range of organic materials can be processed to produce vermicompost with the addition of a little pre-processing. Pre-processing usually involves:

(i) Size reduction: To increase the surface area for microorganisms to attack.
(ii) Mixing: To achieve a suitable structure, moisture content and nutrient balance.
(iii) Addition of a bulking agent: To improve structure, increase surface area and to absorb excess moisture. Earthworms more readily process a mixture of compostable organic materials rather than monostreams of specific waste types. The size of organic materials to be processed by a vermiculture unit should be reduced to enable effective processing by the worm population. The addition of a bulking agent is required to form a feedstock to support problem-free processing.

(iv) Mixed food: Organics (mixed rotten fruits and vegetables, stale bread, meat/poultry residues etc).

(v) Mixed garden materials: This includes lawn clippings, and non-woody plant materials such as stems, leaves and twigs of various plant species. The addition of a bulking agent, such as waste paper or cardboard, is very important when preparing compostable organic materials for processing in a vermiculture unit. Cardboard and paper are carbonaceous materials and absorb excess moisture, increase the porosity and structure of the material and increase the carbon to nitrogen (C:N) ratio. It is the ratio of the weight of organic carbon to total nitrogen within the material. Some organic materials, such as meat and poultry, are rich in nitrogen. If these nitrogen-rich organic materials are processed in a vermiculture unit, carbon is required to be added to achieve a C:N ratio of 20 to 25 parts carbon to every one part of nitrogen.

Carbon can be added to a vermiculture unit as shredded paper or cardboard. These high carbon materials are called bulking agents and are common packaging wastes. The addition of a bulking agent, such as paper or cardboard, not only increases the C:N ratio but also improves the structure and porosity of the material. All these factors produce a material that is more readily processed by the worm population. The omission of compostable organic materials with a bulking agent to increase the C:N ratio may result in the material becoming too dry. The final mixture of organic material amended with a bulking agent and water is called feedstock. Feedstock is the result of blending different components to produce a suitable source of food for the worm population. These factors are important for acceptance of the feedstock by the worm population. Worms need a moist environment, so the material they consume needs to be moist but not too wet. These factors are important for acceptance of the feedstock by the worm population.

4.7. Materials That Cannot Be Processed for Vermicomposting

Vermiculture technology cannot process all categories of compostable organics with the same efficiency. Only a limited range of organic materials can be effectively managed through on-site vermiculture technology. The audit of 'waste' material produced on-site is an essential step in identifying the type of compostable materials. Materials that are very high in nutrients such as seafoods and dairy products are not recommended for vermiculture processing in any significant proportion. These materials can cause problems such as anaerobic conditions that result in the worms' death. Microorganisms break down these high nutrient materials very quickly, resulting in rapid oxygen consumption. This can lead to health and safety issues such as odor production and the attraction of pests. Preparing a suitable feedstock for processing in a vermiculture unit is a crucial step in ensuring a healthy environment for the worm population. Failures of vermiculture units can be attributed to the addition of unsuitable feedstock or excessive quantities of feedstock. Problems that can arise from an unsuitable feedstock include:

(i) *Feedstock too moist*: results in anaerobic (low oxygen) conditions.
(ii) *Feedstock too dry*: not suitable for worm movement and habitation.
(iii) *Feedstock containing high proteinous food* (such as seafood or dairy material): can't be readily processed by the vermiculture unit.
(v) *Feedstock loading rate too high*: results in low oxygen conditions and worm death.

The processing capacity of vermiculture technology refers to the maximum amount of organic material that can be added to a vermiculture unit per unit of time without causing system failure. Vermiculture units have a limit to the amount of organic materials that can be processed over a period of time. If the maximum processing capacity is exceeded, problems such as anaerobic conditions, worm death, odor production, pest attraction and ultimate system failure may arise. The maximum processing capacity of a vermiculture unit depends on the type of materials being fed to it, as worms process different organic materials at different rates. The installation of a vermiculture unit will only be successful if the appropriate compostable material is processed in the unit at an appropriate loading rate. The maximum processing capacity is expressed as the volume of feedstock applied per square meter of bedding surface per week.

4.8. Stocking the Unit with Worms

Worms used in vermiculture units tend to process larger amounts of organic material, reproduce in confined environments and cope well with disturbances (such as feeding and maintenance procedures) when compared with other common species. As many as 418 species of earthworms are known from India; about a dozen have been identified as useful in vermicomposting systems. These represent both exotic and native species.

These species were evaluated based on their:

- Tolerance to a wide range of environmental conditions and fluctuations
- Tolerance to disturbance
- High growth
- High breeding rate

Earthworm species with a short generation time, meaning a relatively short lifespan and rapid growth and reproductive rate, have been identified as the most effective species. In India two exotic species - namely, *Eudrilus eugeniae* and *Eisenia fetida* - and the native species *Perionyx excavatus* are extensively used, and in fact the vermicomposting industry in India is primarily based on these species.

When establishing a vermiculture unit, the appropriate type of worms need to be incorporated in a sufficient quantity to process the organic materials produced on-site. Depending on the feedstock types, a worm application rate of between 10 and 18 kg per meter of bedding surface (10-18 kg/m^2) is recommended as a sufficient rate to quickly establish maximum processing capacity.

Introducing the worm stock to a vermiculture unit should be performed in the morning. This will ensure that the worm population does not exit the unit, as daylight will deter the worms from escaping. Alternatively, a bright light can be used to encourage the worm stock to burrow into the new environment. Sprinkle the worm stock carefully over the surface of the vermiculture unit. Then the worms will quickly burrow into the bedding material. Ensure the bedding material is moist but not too wet. A period of acclimatization is necessary for the worm stock when introduced to this new environment. Refrain from applying feedstock for a week after introducing the worms to allow them to settle. Rectifying a failed system requires much more work than managing a system effectively.

4.9. Size and Site Selection for Installation of Vermiculture Unit

Selection of the type and size of vermiculture unit varies from site to site depending on a number of factors. These include:

i. Type of materials to be processed
ii. Cost of production
iii. Purpose of the vermiculture unit
iv. Availability of complementary materials
v. Vermiculture processing capacity for the organic material
vi. Availability of space and other site-specific constraints.

In order to select the size of vermiculture unit a 'waste' audit needs to be conducted to determine the quantity and type of materials produced. The size of the vermiculture unit depends on available organic materials. If the vermiculture unit is too small, the process is likely to fail. When installing the vermiculture unit, careful consideration should also be given to the site or location of the vermiculture unit and ancillary equipment. Determining the most suitable location of an on-site mid-scale vermiculture unit is dependent on a number of factors. These include:

i. Accessibility
ii. Maintaining climatic conditions (shading and controlling temperature and moisture)
iii. Security measures
iv. Proximity to neighborhood
v. Area for storage of materials and feedstock preparation
vi. Noise and odor considerations
vii. Leachate
viii. Pest exclusion
ix. Availability of services such as water and power supply

Site selection for an on-site vermiculture installation is important both for efficiency and building materials. The worm activity is dependent upon environmental conditions including temperature and moisture. Proper placement and management ensures the effective operation of the vermiculture unit. Effective monitoring and management of the vermiculture unit is essential for its efficient operation. Special attention is needed for:

i. Worm activity
ii. Feedstock accumulation
iii. Oxygen
iv. Temperature
v. Moisture content

Indicators of Stress in Functioning of Worms

Some factors within the system may indicate system success or failure and these should be regularly investigated as indicators of system health. The hierarchy of system performance indicators is as follows:

Worm activity: This is the first indication that a system is stressed. Worms will attempt to escape the vermiculture unit if conditions are unsuitable for habitation.

i. Feedstock accumulation: Feedstock will accumulate if the worms are no longer processing it and are attempting to escape. The feedstock will accumulate and cause the conditions to be even more unsuitable for worm habitation.
ii. Oxygen: If feedstock accumulates, oxygen levels will decrease and cause anaerobic conditions. This will add to the uninhabitable conditions of the unit.
iii. Temperature: Finally, the temperature will rise within the unit as feedstock accumulates and decomposes.

4.11. Multiplication of Worms on a Large Scale

There are many creative ways in which a farmer can manage for earthworms' *in situ* conditions. A first step might be to determine what earthworm ecotypes are present, and how abundant they are. Endogeic species are most commonly found. These are useful, but a mixed community including anecic species as well would be even more beneficial, especially for the incorporation of surface matter. Direct inoculation is one possible method, but transferring blocks of soil (one-cubic foot each) from an area with a large earthworm population into farm soil might work better. One management idea for introducing desired species is to set aside a small area of land on a farm to be managed exclusively as an earthworm reservoir. If needed, the soil could be limed to bring it near pH 7, fertilized, and a cover crop established and cut periodically to provide organic mulch as food and physical cover. In this area a community of the desired species could be established and built up. From this reservoir blocks of worms could periodically be taken and introduced into the field.

This is a long-term process for establishing earthworms, and would only be successful if ample organic matter was supplied to the soil where earthworms were being introduced, and also if physical and chemical disturbances of the soil were minimized. Organically managed perennial crops would be ideal for this method. The inoculation of epigeic earthworm at 50 per kg mixture of cow dung and dried leaves (1:1) in controlled conditions also helps the fast multiplication of earthworms.

4.12. Earthworm Species Used for Vermicomposting in India

(i) Exotic

- *Pontoscolex corethrurus* Muller
- *Eisenia fetida* Savigny
- *Eudrilus eugeniae* Kinberg
- *Metaphire posthuma* Vaillant
- *Dichogaster boloni* Michaelsen
- *Polypheretima elongate* Perrier

(ii) Native

- *Perionyx excavatus* Perrier
- *Perionyx sansibaricus* Michaelsen
- *Perionyx ceylanensis* Michaelsen
- *Megascolex cochinensis* Stephenson
- *Drawida willsi* Michaelsen
- *Drawida nepalensis* Michaelsen
- *Lampito mauritii* Kinberg
- *Haplochaetella suctoria* Stephenson
- *Haplochaetella khandalensis* Stephenson

- *Octochaetona surensis* Michaelsen
- *Moniligaster perrieri* Michaelsen
- *Megascolex insignis* Michaelsen
- *Drawida robusta* Bourne

4.13. Significant Properties of Vermicompost

(i) High levels of bio-available nutrients for plants: Vermicompost contains most nutrients in plant-available forms such as nitrates (N), phosphates (P), soluble potassium (K), and magnesium (Mg) and 'exchangeable' phosphorus (P) and calcium (Ca) (Edwards 1996, 1998, 2004). Vermicomposts have a large particulate surface area that provides many micro-sites for microbial activities and for the strong retention of nutrients (Arancon et al., 2004, 2006).

(ii) High level of beneficial soil microorganisms promoting plant growth: Vermicomposts are rich in 'microbial populations and diversity', particularly fungi, bacteria and actinomycetes (Brown, 1995; Chaoni et al., 2003; Sadhale, 1996; Singh, 2009; Sukumaron, 2008). Teotia et al. (1950) and also Parle (1963) reported the bacterial count of 32 million per gram in fresh vermicast compared to 6-9 million per gram in the surrounding soil. Scheu (1989) reported an increase of 90% in respiration rate in fresh vermicast indicating a corresponding increase in the microbial population. Suhane et al. (2008) found that the total bacterial count was more than 10^{10} per gram of vermicompost. It included Actinomycetes, Azotobacter, Rhizobium, Nitrobacter and phosphate solubilizing bacteria which ranged from 102-106 per gm of vermicompost. The PSB has a very significant role in making the essential nutrient phosphorus (P) 'bio-available' for plant growth promotion (Rodríguez et al., 2004).

Although phosphates are available in soils in rock forms, they are not available to plant roots unless solubilized. Pramanik et al. (2007) studied the microbial population in vermicompost prepared from cow dung and municipal solid wastes (MSW) as substrates (raw materials) and found that it was in highest abundance in cow dung vermicompost. The total bacterial count was 73×10^8, the cellulolytic fungi were 59×10^6 and the nitrogen-fixing bacteria were 18×10^3. It was least in vermicompost obtained from MSW. The total bacterial count was 16×10^8, the cellulolytic fungi were 21×10^6 and the nitrogen-fixing bacteria were 5×10^3. The application of lime in the substrate enhanced the population of all the abovementioned microbes irrespective of the substrates used for vermicomposting. Plant growth promoting bacteria (PGPB) directly stimulates growth by nitrogen (N) fixation, solubilization of nutrients, production of growth hormones such as 1-aminocyclopropane-1-carboxylate (ACC) deaminase and indirectly by antagonising pathogenic fungi by production of siderophores, chitinase, ß-1,3-glucanase, antibiotics, fluorescent pigments and cyanide (Han et al., 2005).

There is also a substantial body of evidence to demonstrate that microbes, including bacteria, fungi, actinomycetes, yeasts and algae, also produce 'plant growth regulators' (PGRs) such as auxins, gibberellins, cytokinins, ethylene and ascorbic acids in appreciable quantities and, as their population is significantly boosted by earthworms, large quantities of PGPRs are available in vermicompost (Frankenberger and Arshad, 1995).

(iii) Rich in growth hormones: Research shows that vermicompost further stimulates plant growth even when plants are already receiving 'optimal nutrition'. Vermicompost has consistently improved seed germination, enhanced seedling growth and development and increased plant productivity much more than would be possible from the mere conversion of mineral nutrients into plant-available forms. Arancon et al. (2003) found that maximum benefit from vermicompost is obtained when it constitutes between 10 to 40% of the growing medium. Neilson (Neilson, 1991; 1965) and Tomati et al. (1988) have also reported that vermicompost contained growth promoting hormone 'auxins', 'cytokinins' and flowering hormone 'gibberellins' secreted by earthworms. It was demonstrated by Grappelli et al. (1985) and Tomati (1983, 1987 & 1988) that the growth of ornamental plants after adding aqueous extracts from vermicompost showed similar growth patterns as with the addition of auxins, gibberellins and cytokinins through the soil.

(iv) Rich in humic acids: Atiyeh et al. (2000) speculated that the growth responses of plants from vermicompost appear more like '*hormone-induced activity*' associated with the high levels of humic acids and humates in vermicompost rather than being boosted by high levels of plant-available nutrients. This was also indicated by Canellas et al. (2000) who found that humic acids isolated from vermicompost enhanced root elongation and formation of lateral roots in maize roots. Pramanik et al. (2007) also reported that humic acids enhanced 'nutrient uptake' by the plants by increasing the permeability of the root cell membrane, stimulating root growth and increasing the proliferation of 'root hairs'.

(v) Vermicompost is free of pathogens: Nair et al. (2007) found that 21 days of a combination of thermocomposting and vermicomposting produced compost with an acceptable C:N ratio and good homogenous consistency of a fertilizer. The study also indicated that vermicomposting leads to a greater reduction of pathogens after 3 months of storage, whereas the samples which were subjected to only thermofilic composting retained higher levels of pathogens even after 3 months.

(vi) Vermicompost is free of toxic chemicals: Several studies have found that earthworms effectively bio-accumulate or biodegrade several organic and inorganic chemicals including 'heavy metals', 'organochlorine pesticide' and 'polycyclic aromatic hydrocarbons' (PAHs) in the form of residues in the medium in which it inhabits.

(vii) Vermicompost protects plants against various pests and diseases: There has been considerable evidence in recent years regarding the ability of vermicompost to protect plants against various pests and diseases either by suppressing or repelling them or by inducing biological resistance in plants to fight them or by killing them through pesticidal action (Al-Dahmani et al., 2003).

(viii) Induce biological resistance in plants: Vermicompost contains some antibiotics and actinomycetes which help in increasing the '*power of biological resistance'* among the crop plants against pest and diseases. Pesticide spray was significantly reduced where earthworms and vermicompost were used in agriculture.

(ix) Repel crop pests: There seems to be strong evidence that worms' vermicastings sometimes repel hard-bodied pests (Arancon, 2004). Edwards and Arancon (2004) report a statistically significant decrease in arthropods (aphids, buds, mealy bugs, spider mites) populations and subsequent reduction in plant damage, in tomato, pepper and cabbage trials with 20% and 40% vermicompost additions. George Hahn, performing commercial vermicomposting in California, claims that his product repels many different insects' pests. His explanation is that this is due to the production of '*chitinase*' enzymes by worms which break down the chitin in the insect's exoskeleton (Munroe, 2007).

(x) Suppress plant disease: Edwards and Arancon (2004) found that the use of vermicompost in crops inhibited soil-born fungal diseases. They also found statistically significant suppression of plant-parasitic nematodes in field trials with peppers, tomatoes, strawberries and grapes. The scientific explanation behind this concept is that high levels of agronomically beneficial microbial population in vermicompost protect plants by out-competing with plant pathogens for available food resources i.e. by starving them and also by blocking their excess to plant roots by occupying all the available sites. This concept is based on '*soil-food web*' studies pioneered by Dr Elaine Ingham of Corvallis, Oregon, U.S. (http://www.soilfoodweb.com). Edwards and Arancon (2004) reported the agronomic effects of small applications of commercially produced vermicompost on attacks by fungus *Pythium* on cucumber, *Rhizoctonia* on radishes in the greenhouse, by *Verticillium* on strawberries and by *Phomposis* and *Sphaerotheca fulginae* on grapes in the field. In all these experiments vermicompost applications suppressed the incidence of the disease significantly. They also found that the ability of pathogen suppression disappeared when the vermicompost was sterilized, convincingly indicating that the biological mechanism of disease suppression involved was '*microbial antagonism*'. Szczech et al. (1993), Orlikowski (1999), Rodriguez et al. (2000) and Zaller (2006) also found that the aqueous extracts of vermicomposts depress soil-borne pathogens and pests. They found in their field experiment that only half as many plants of tomatoes sprayed with aqueous extract of vermicompost were infected with *Phytophthora infestans* (that cause 'late-blight' disease) compared to the control plants. Vermicompost benefits soil in the following ways:

(i) Improving its physical structure
(ii) Enriching soil with microorganisms, adding plant hormones such as auxins and gibberellic acid, and adding enzymes such as phosphatase and cellulase.
(iii) Attracting deep-burrowing earthworms already present in the soil
(iv) Improving water-holding capacity of soil
(v) Enhancing germination, plant growth, and crop yield
(vi) Improving root growth and structure
(vii) Vermicompost is an ecofriendly natural fertilizer prepared from biodegradable organic wastes and is thus free from chemical inputs
(viii) It does not have any adverse effect on soil, plant or environment
(ix) It improves soil aeration, texture and tilth, and thereby reduces soil compaction
(x) It improves the water retention capacity of soil as it is rich in organic matter content
(ix) It promotes better root growth and nutrient absorption.
(x) It improves the nutrient status of soil - both macro-nutrients and micro-nutrients.

4.14. Factors Determining the Nutritional Quality of Vermicompost

The nutritional quality of vermicompost is determined primarily by the type of the substrate (raw materials) and species of earthworms used for composting, along with microbial inoculants, liming, aeration, humidity, pH and temperature. Cattle dung has been found to yield most nutritive vermicompost when composted by *Eisenia fetida*. Pramanik et al. (2007) found that the application of lime at 5 gm/kg of substrate and 'microbial inoculation' by suitable 'cellulolytic', 'lignolytic' and 'N-fixing' strains of microbes not only enhances the rate of vermicomposting but also results in nutritionally better vermicompost with greater enzymatic (phosphatase and urease) activities. Kumar et al. (2010) found that inoculation with N-fixing bacteria significantly increased the 'nitrogen' (N) content of the vermicompost. Liming generally enhances earthworm activities as well as the microbial population.

The earthworms ingest microbes into their guts and proliferate the population of microbes to several times in their excreta (vermicasts). It is therefore advantageous to use beneficial microbial inoculants whose population is rapidly increased for rapid composting and also better compost quality. Pramanik et al. (2007) studied the vermicomposting of four substrates - namely, cow dung, grass, aquatic weeds and municipal solid wastes (MSW) - to determine the nutritional status and enzymatic activities of the resulting vermicomposts in terms of increase in total nitrogen (N), total phosphorus (P) and potassium (K), humic acid contents and phosphatase activity.

Mulching also affects the functioning of vermicompost to keep it moist and allows the worm cocoons to germinate faster into baby worms and the beneficial microbes to multiply and act faster. Apparently, both the earthworms and their excreta (vermicasts) play a combined role in growth promotion. Worms and microbes secrete growth promoting plant hormones, gibberlins, auxins and cytokinins, to help mineralize the nutrients and make them 'bioavailable'. In a glasshouse trial, Buckerfield et al. (1999) found that the stimulatory effect of vermicompost on plant growth was apparently destroyed when it was sterilized.

4.15. Factors Affecting the Earthworm's Growth

(i) Pest: The major earthworm predator is the mole. This voracious insect predator loves to dine on white grubs and any earthworm it can find. Grubs, attached to the root from which they gain their food, can't escape, but the earthworm can feel the vibrations of the mole digging and quickly try to flee. The mole's own digging conceals the noise of the earthworm fleeing, so the star-nosed mole has developed a unique method to find earthworms. It uses its fussy-looking nose to detect the faint electrical fields that earthworms (and some other insects) radiate. Not only does this mole detect and find an earthworm, it also knows how to bite it so it is paralyzed but does not die. The mole stores the live worm along the burrow as food for dining at leisure. Another predator, usually found in the northeast, is the carnivorous snail; they also love to eat earthworms. Garden centipedes can be a problem in the worm bin, especially outside. These predators should be destroyed. Overly wet beds can also attract the earthworm mite, which causes the worms to stop eating. Leeches are also known to attack vermeries.

Nematodes are found emerging from the cocoons of *Eudrilus eugeniae* which affect their viability. Incidence of parasitism increases with the age of the host. There are records of different species of parasitic nematodes from the coelomic cavity of the earthworms. Different species of *Capillaria annulata*, a rotifer *Alberitia vermicularis*, have been isolated from *Allolobophora calliginosa*. Ciliate and sporozoan parasites have been isolated from the body fluid and various tissues of earthworms.

(ii) Tilling and Earthworm Population: Tilling the soils does reduce the earthworm population; not because it kills or disturbs them, but because tilling increases aeration in the soil, and this oxygen quickly reduces the organic matter that the earthworm uses as food. Mulching with green matter helps to provide food to earthworms to replenish what is lost in tilling. The population of earthworms, in the north, follows a different cycle from that of most garden fauna. The population of adults is highest in the spring, and decreases in the dry summer months, followed by an increase of young in the winter. For a high number of earthworms in the spring, it's important to protect the young and the eggs over winter.

(iii) Earthworm and Water logging: Earthworms come out of their burrows during the rain season to avoid drowning. Worms have no lungs; they take their oxygen directly through the skin, either from air or from water. In fact, rather than fear water, they love it. Its drying out they fear and dry soil kills them. When it rains, they come to the surface because it's easier to find a mate in the flat open ground than the bin or the three-dimensional burrows. The wet ground allows them to move without fear of drying out.

(iv) Overloading of Feed Material: Overloading the bin with food wastes can result in foul odors. Overly moist food waste and bedding also cause bad odors. To relieve this problem, fluff up the bedding to add air and check the drainage holes. As a general rule, keep the bedding material moist, but never soggy. Make sure the food waste is buried properly in the bedding. Exposed food wastes can attract fruit flies, house flies, and other pests. Keeping the bin covered with straw or moist burlap also deters these pests.

4.16. General Problems in Production of Vermicomposting

It is unlikely that the worm bin will have an unpleasant odor. If it does, there are a number of possible causes and steps you can take to remedy the problem.

(i) *The bin is overloaded with too much food waste.* Solution: Don't add any more food for a week or two.

(ii) *The bedding is too wet and compact.* Solution: (a) Gently stir the entire contents to allow more air in and stop adding food waste for a week or so. Make sure that food waste is still buried; (b) The lid can be removed or left slightly to allow the contents to dry out.

(iii) *The bin is too acidic.* Solution: Add some calcium carbonate and cut down the amount of citrus and other acidic food waste.

(iv) *Infestation of Mites.*Solution: (a) More watering than usual,*but do not flood the worm beds*. Mites will move to the surface, and worms will stay below the surface. Use a hand-held propane torch to scorch the top of the bed and kill the mites. This procedure may be repeated several times, at three-day intervals, if needed;(b) Light dusting of sulphur may be used to kill the mites. After soaking the worm bed with water and causing the mites to surface, apply a rate of 1/16 ounce of sulphur per square foot of bed surface. Sulphur should not harm the worms, but at times it may increase the acidity of the bed and reduce earthworm populations.

4.17. Troubleshooting in *Vermicomposting**

Sl.No.	*Problem*	*Probable Cause*	Solution(s)
1.	Worms are dying or trying to escape	Bed too wet	Add more bedding
		Bed too dry	Moisten bedding
		Bedding used up	Harvest bin
2.	Bin smells bad	Not enough air	Leave lid off or drill more ventilation holes
		Too much food	Do not feed for 1-2 weeks
		Non-compostable waste present	Remove non-compostable waste
3.	Bin attracts flies	Food scraps exposed	Bury food in bedding
		Rotten food	Avoid putting rotten food in bin
4.	Bin smells bad	Rotten food	Stop feeding for 2 weeks
		Too much food, especially citrus	Bury food completely
		Bin too wet	Drill holes in bin
5.	Worms are dying	Bin too wet	Mix in dry bedding; drill holes in bin
		Bin too dry	Maintain moisture
		Not enough food	Add more bedding and food scraps
		Extreme temperature	Move bin where temp between 55 and77°F
6.	Worms crawling away	Bin conditions not right	See solutions above; Leave lid off and worms will burrow back into bedding
7.	Mold forming	Conditions too acidic	Cut back on citrus fruits
8.	Bedding drying out	Too much ventilation	Dampen bedding; keep lip on

(Continued)

Sl.No.	Problem	Probable Cause	Solution(s)
9.	Water	Poor ventilation	Leave lid off for a couple of days; add dry bedding
10.	Collecting in bottom	Feeding too much watery scraps	Cut back food scraps with high water content

* Source: The complete technology book on Vermiculture and Vermicompost, National Institute of Industrial Research, 2004.

Chapter 5

EARTHWORMS AND THEIR USE IN VERMITECHNOLOGY

The earthworm (intestine of earth) is a very important component for the maintenance of soil fertility and the nutrient cycle. Darwin (1881) for the first time reported the role of earthworms in increasing soil fertility. Earthworms are able to convert even organic parts of municipal solids into rich manure. They are also able to deodorize the composting process. Vermitechnology is both an ecologically and economically viable process and can be adopted by any ordinary person. Thus, these tubular creatures are now largely exploited by man for economic returns. They also have a role in ecological soil management and organic waste recycling.

Earthworms can play a variety of important roles in agro ecosystems. Their feeding and burrowing activities incorporate organic residues and amendments into the soil, enhancing decomposition, humus formation, nutrient cycling and soil structural development. Earthworm burrows persist as macropores which provide low resistance channels for root growth, water infiltration and gas exchange. The quality, quantity and placement of organic matter are a main determinant of earthworm abundance and activity in agricultural soils; they may be disturbed due to soil tillage, cultivation, and the use of pesticides.

Earthworms harbor millions of *'nitrogen-fixing' and 'decomposer microbes'* in their gut. They have to necessarily feed upon microbes, particularly fungi, to meet their protein/nitrogen requirement essential for growth and reproduction. Earthworms also produce a huge amount of intestinal mucus composed of glyco-proteins and small glucosidic and protein molecules. The microbes entering the gut of worms consume all these nitrogenous compounds of the mucus, which largely increase their activity, which in turn enables them to contribute enzymes in the digestive process of earthworms. The microbes not only mineralize the complex substances into plant-available form but also synthesize a whole series of biologically active substances. The worm's body contains 65% protein (70-80% high quality 'lysine rich protein' on a dry weight basis), 14% fats, 14% carbohydrates and 3% ash. Earthworms act as an aerator, grinder, crusher, chemical degrader and a biological stimulator wherever they inhabit.

As many as 418 species of earthworms are known from India which is about 10.5 per cent of the total known global earthworm diversity, estimated at 4000 species. Out of these vast earthworm resources only half a dozen species are frequently used in India for vermiculture and vermicomposting. These are *Eudrilus eugeniae, Eisenia fetida* (including its supposed hybrid, *Eisenia anderei), Perionyx excavatus, Lampito mauritii, Polypheretima elongata* and *Pontoscolex corethrurus*. Among these, only *Perionyx excavatus* and *Lampito maurtii* are natives of India and the rest are exotic species. The vermicomposting industry here is primarily based on the African *Eudrilus eugeniae* and the European *Eisenia fetida*. The *Eisenia fetida* is utilized all over the country in both warm and cold climates while the *Eudrilus eugeniae* is widely used in southern parts of the country with a warmer climate. Researchers are exploring more native species that can be used in vermicomposting.

5.1. Origin of Earthworms

Earthworms are among the most ancient of terrestrial animal groups, their ancestors possibly emerging in the late Precambrian period some 650-570 Ma (mega-annum = a million years) ago (Valentine, 1980; Bouché, 1983; Lee, 1985). Indeed, the lower boundary of the Cambrian 543 Ma is now defined by the common occurrence of horizontal worm-like burrow trace fossils; even earlier dates in the Proterozoic era 1 billion years ago have been proposed by Seilacher et al. (1998). Lee (1985) cites reports by Glaessner et al. (1969) of marine annelid fossils in South Australian Precambrian sediments (e.g. the Ediacara Hills Assemblages that now date the Ediacaran period extending back as far as 635 Ma), and segmented worms have also been identified from basal Cambrian strata in China and Siberia (Bengtson & Zhao, 1997), and from the upper Ordovician and upper Silurian periods in the USA and the UK (Bather, 1920; Edwards & Lofty, 1977). Michaelsen (1903)'s opinion initially was that earthworms arose not later than the upper Triassic period, but later Michaelsen (1922) thought 'real' earthworms appeared in the Carboniferous era. Fossil embryos and cocoons were found in recent and Mesozoic layers (Schwert, 1979; Piearce et al., 1990; 1992; Manum & Bose, 1991), and Late Tertiary and Quaternary beds yielded the calciferous glands of *L. terristris* that were previously named *Arion kinkeli* and *A. hochheimensis* fossils (Gates, 1972). Trace fossils of worm burrows or casts are known from Grinshill sandstone that also bares the remains of reptilian tetrapods, has trace fossils of worm casts and "a meandering 'worm' trail about 15 mm wide" (Bengston & Zhao, 1997) from the Triassic period 220M years ago that preceded the Jurassic age of dinosaurs (http://www.darwincountry.org/category.php3?trail=2997).

'Bioneers' in the colonization and preparation of the land, these ancestral forms pre-date the invasion of fungi, land plants, insects (400-500 Ma), dinosaurs and mammals (200 Ma), and occur well before the descendents of these early worms – the hominids – arose (4Ma) (Launer, 2006). Earthworms now are ubiquitous in all but the driest of regions and the present-day world distributions of the 20 families have been explained in terms of Wegener's hypothesis of continental drift (Michaelsen, 1922; Lee, 1985; Sims, 1980). Wegener L. Alfred (1880-1930) formulated his ideas in 1911 and published them in 1915 after recovering in a military hospital from injuries sustained on the Western Front in the Great War of 1914-1918.

Earthworms have left no known Pre-Neogene Fossils of any kind, save possibly ichno fossils of a form consistent with the modern earthworm – created structures. Thus, the origins of earthworms are uncertain in time and space, but their position on the phylogenetic tree of Annelida: Clitellata is probably as a sister group to the Enchytraeidae (James, 2004). Ancestral earthworms are supposedly furnished with an octogonadal battery comprising four testes, a pair each in 10 and 11, and four ovaries, a pair each in 12 and 13 (Gates, 1982). African *A. mutivesiculatus* and American *L. multivesiculatus* have nine pairs of testes in 10-18, and ten pairs of testes in 12-21, respectively.

Earthworms of the world belong to not more than a dozen families. According to Wallwork (1984), this is a remarkably small number for a group of invertebrates that has been around for a very long period, and has successfully spread to soil throughout the world. Possibly, these organisms have not been able to exploit diverse food niches in an evolutionary sense. Sims (1980) assumed that the ancestors of present day earthworms were widespread in the undivided pale continent of Pangaea, which existed during the Paleozoic.

5.2. Biology and Ecology of Worms

Accounts of oligochaete biology, morphology, and terminology are found in Benham (1890 – historically interesting but a bit outdated), Stephenson (1930 – still valid), Lee (1959), Gates (1972), Julka (1988), Lee (1985), Sims and Gerard (1985; 1999) correct, mostly), Edwards and Bohlen (1996, most basics) and Csuzdi and Szlavecz (2003 for Lumbricidae). An elementary account of earthworms, but with several superceded names, can be found online (Wikipedia). However, it may be useful now to provide an introduction to the subject for the benefit of students/ecologists intending to enter this field of study, and for the practical application of findings from agro-ecological field studies, whether conducted in India or elsewhere in the world.

Earthworm communities may comprise four components (Blakemore, 1999): *resident natives* that are often highly endemic (i.e. indigenous and restricted to a certain area); introduced *exotics* that tend to be more widespread; *translocated natives* i.e. endemic species that have been transported (by humans) within a bioregion; and "*neoendemics*" i.e. those species that have sojourned for sufficient time, following introduction, to have undergone speciation in a new region (or are currently unknown or extinct in their places of origin) – examples described as possibly neoendemic are *Lumbricus badensis* (Michaelsen, 1907) in southern Germany, *Begemius spp.* in northern Australia, several *Microscolex* on subantarctic islands, *Dichogaster* spp. in the Pacific, and several '*Pheretima*' species in the Solomons (Gates, 1958). Many of the common peregrine exotics are known to have parthenogenetic morphs so, in principle, just a single specimen or cocoon is needed for establishment in a new area (Gates, 1972; Sims & Gerard, 1985, 1999).

5.3. Earthworm Population Dynamics

In *The Origin of Species* Charles Darwin (1859) quoted:

> "Throw up a handful of feathers, and all must fall to the ground according to definite laws; but how simple is this problem compared to the action and reaction of the innumerable plants and animals which have determined, in the course of centuries, the proportional numbers and kinds..."

Populations are usually estimated by sampling, hand sorting, and counting the number of worms (just the 'heads') per unit area. In some situations, the amount of mole activity or number of song birds may give rough estimations of earthworm abundance, as indeed do the number of surface worm casts or verdant plant growth. Adequate sampling requires good ecological methods such as sufficiently large quadrate size and sample numbers, with consideration of depth, season, immature/cocoons, "edge effects", etc. And good taxonomic skills/guides are needed to identify the community component – this includes a thorough search of the historical scientific literature to find earlier species records.

Lee (1985) summarized abundance and diversity data from regional habitats. High diversity has been recorded from some sites; for example, from a total of about 40 species identified during three years' observations around at Samford in south-east Queensland (Blakemore 1994; 1997) 24 species (5 exotics and 16 natives plus 3 microdriles) were collected in one week at Lake Pedder, Tasmanian Wilderness World Heritage Area (Blakemore, 2000; 2010) and fifteen species (10 exotics and 4 natives plus enchytraeids) were found after a couple of days' survey on a 45 ha land at Budderoo in the Southern Highlands of New South Wales, Australia (Blakemore, 2001). Meanwhile, the average Australian backyard garden can support a varied community, often of a dozen or more species, both known and new (Blakemore, 1994; 1997).

In Taiwan, a quadrate survey from coastal plain to summit at 3,200m yielded 34 species consisting of 19 native and 15 exotic species (Tsai et al., 2004). A survey of Tam Dao National Park, Vietnam in October 2005 of natural and cultivated lands found 38 earthworm sp. belonging to 5 genera, 4 families (Blakmore, 2000). Thus, the biodiversity and distributions of worms may act as indicators to monitor soil pollution/health.

The earthworm possesses limited means of active migration, but passive means have distributed some species over wide areas. Most of them cannot survive marine environments which are, therefore, effective barriers for extension of their natural distribution. Endogenic species have, in general, a poor capacity to migrate (Bouche, 1983). Some of them are very sensitive to desiccation and do not leave the soil. Anecic worms are able to leave their burrows under certain conditions and colonize a few meters per annum (Mazaud and Bouche, 1980). Epigeic forms are most mobile and travel some distances (sometimes on a mass scale) in search of food and shelter. Man is also responsible for transporting a few species for culture and in soil around the roots of exotic plants. It is well known that anthropochrous species are able to colonize in disturbed environments which may be created by deforestation, intensive cultivation in a new area, pesticides, waste disposal etc. Natural destruction of earthworm habitats may be due to glaciations, tectonic disturbance and changes in sea levels. The presence of anthropochorous or peregrine species in an area sometimes obscures the distributional pattern of earthworms.

Coupland and McDonald (2008) reported high populations of *Pontodrilus litoralis* of 750-4,875 m^{-2} under wrack material on arid beaches in Western Australia and calculated that earthworms could consume approximately 19-31 kg m^{-2} yr^{-1} organic material there. One of the highest enchytraeid mean abundance reports was of ca. 200,000m^{-2} at Moor House, England (Lavelle & Spain, 2001).

More usually, populations of earthworms in rich organic soils range up to 1,000~2,000m^{-2} with biomass as high as 5,000kg ha^{-1} (5t ha^{-1}), and whereas smaller worms and enchytreids may number 290,000m^{-2}, nematodes number in the millions per square meter, there can be up to 250,000m^{-2} oribatid mites, and innumerable protozoans can equal or exceed the biomass of the earthworms. Naturally, microbes are the most abundant and diverse in most habitats.

Table 5.1. The total animal biomass of soil fauna

Group	Individuals m^{-2} (approx.)	Biomass g m^{-2}
Protozoa	10^{10}	6-30
Nematoda	10^{6}	1-30
Earthworms	50-500	20-500
Acari	10^{4}	0.2-4
Collembola	10^{4}	0.2-4
Insecta larvae	50-500	4.5
Myriapoda/Diplopoda	20-700	0.5-12.5
Myriapoda/Chilopoda	100-400	1-10
Isopoda	Up to 1800	Up to 4

The total animal biomass of a typically fertile soil (not counting algae or fungi) is estimated at 2.5t ha^{-1} for which the earthworms make up the bulk of this, as shown in the following table of soil fauna populations typical of temperate regions (after Gobat et al., 2004, Table 5.1). Thus, the top 10 cm of fertile topsoil can contain 200kg m^{-2} of organic carbon locked up in these living soil organisms and in the humic matter (plus glomalin) they consume and recycle. Organic farming can substantially preserve high levels of earthworm abundance and (bio) diversity and of organic carbon in managed soils, as was clearly shown in the Haughley Experimental site (Blakemore, 1981, 2000; Blakemore and Temple–Smith, 1996).

5.4. GLOBAL DIVERSITY OF EARTHWORMS

Michaelsen (1900) described 1,200 species of Oligochaetes. His taxonomic efforts have overwhelmed and polarized researchers either to the aquatic microdriles, or to the terrestrial megadriles (the 'true' earthworms) [Benham (1890); Michaelsen (1903)]. Updated indices of most names are provided in a *Nomenclatura Oligochatologica* series by Reynolds and Cook (1976; 1981; 1989; 1993) that lists all names regardless of synonymies. As of December 2005, these nominal taxa approximated as: 38 families; 804 genera; 61 sub-genera, and ~ 8,500 species divided roughly 50:50 into aquatic microdriles plus enchytraeids: terrestrial megadriles (extrapolated from Wetzel, 2005). The ratio of species to genera to families is approximately 225:21:1 (with almost half the relative number of families of Polychaeta).

Reynolds (1994) described 8000 Oligochaeta, with just 3,600 terrestrial megadriles [presumably ~4,400 microdriles] and, from previous trends, an estimate of ~50 years to double the number of described species to 7,200 spp – these figures were quoted by Fragoso et al. (2003) and stated:

> "The annelids were among the first animals to evolve on the face of the earth, probably appearing in the Edicarian Period (Paleozoic Era), >600 million years ago. It is estimated that there are today over 7000 species of terrestrial oligochaetes, of which we know only about 3600".

A Zoobank total for Oligochaeta (microdriles + megadriles with multiple entries for some species) is 12,212 species, genera and family names while an online catalogue (http://earthworms.elte.hu/) by Dr Csaba Csuzdi lists 5,900 megadrile spp to 2008, and at least 200 names can be added from recent work in Tasmania and Asia (Blakemore, 2000; 2008; 2010), so that a total is 6,100 megadriles. And, if we add the 3,600-4,400 microdriles determined from Reynolds (1994) and Wetzel (2005), this would give a *total Oligochaeta of 9,700-10,500 taxa*, but an unknown number are synonyms, arranged in 804 genera (+ 61 unnecessary sub-genera) and 38 families. Blakemore compiled several checklists: e.g. for Holarctic Lumbricidae (670 valid spp from a total of around 1,150 names), and Oriental pheretimoids (~940 valid spp from 1,400 nominal taxa). These most current datasets still have many *species incertae sedis* or *species inquirendae* that require resolution from inspection of types. Yet probably less than a half or a quarter of earthworms from around the world are known.

For earthworms as a whole, a conservative estimate is for the description of a further 3,000 taxa in the next 10 years with 46 full-time taxonomists, or 46 years with 10 taxonomists (Fragoso et al., 2003). Such an effort would raise the numbers of megadriles (including Moniligastridae) from the current total (estimated above at 5,900) to 8,900 spp. and thus total Oligochaeta (microdriles + megadriles) to >12,000 taxa.

In comparison, the polychaete fauna of Asia comprises of some 67 families, 420 genera and 1,140 species and a comparable estimate of total marine Polychaeta is of about 13,000 names (Zoobank lists *21260* names), although only 8,000 of these were considered reasonable species and these are in 1,000 genera and 82 (http://www.ea.gov.au /biodiversity/ abrs/online-resources/polikey/ index.html#history). Their ratio spp.: genera: families = 160:12:1, or perhaps 100:12:1. But this is mostly a product of the much greater number of full-time taxonomists working on this group [about 300 workers are listed on the Polychaete Researchers Online (PRO) website - about 10 times greater than the number of Oligochaeta taxonomists worldwide]. With the same resources, many more earthworm species could be unearthed and ecologies studied; however, numerous synonymies still exist for those already named, so the true total figures are uncertain.

Lee (1959) published the epitome of eco-taxonomy for the earthworms of New Zealand (192 species + key), recently updated by Blakemore in Lee et al., (2000) and Blakemore (2004; 2007; 2010) to 212 spp (25 exotics) [http://soil-bugs.massey.ac.nz/ oligochaeta.php]. In Asia Stephenson (1923), Julka (1988) Julka and Paliwal (2005) described Oligochaeta (418 megadriles, with a current total of 505 species of India and adjacent countries, except Myanmar) [www.zsienvis.in/Biodiversity/Fauna/17.%20Aannelida.doc]. Gates (1958) reviewed lumbricids exotic to the Orient; Gates (1972) listed 174 Myanmar taxa; Easton (1981) and Blakemore (2003) described earthworms of Japan (82 valid species) including 27 species (17 exotic). Tsai et al. (2000) and Blakemore (2004; 2007) summarized species known from Asia: Taiwan 80 species (33 exotic). Mainland China had 105 species (23 exotic) but a current tally is >240 (Blakemore, 2007) – this includes Hainan that has 32 species (8 exotic) but excludes Taiwan. Blakemore (2004; 2007; 2010) described 96 species (23 exotic) from Korea. Thai and Samphon (1991) described 68 spp from Laos, and Thai and Do (1989) summarized 8 from the Cambodian plains.

Thai et al. (2000) listed 195 species from Vietnam. All of these Asian regional reviews have been revised and updated taxonomically and geographically by Blakemore (2004; 2007).

European earthworm faunas are relatively well reported [an EU *Fauna Europaea* (www.faunaeur.org/full_results.php?id=11870 Jan, 2007) lists 472 (sub)species for Suborder Lumbricina], but several taxa are obviously American; and the EUNIS database has 505 megadriles with 463 Lumbricidae that includes many synonyms (eunis. eea.europa.eu/ species-taxonomy.jsp?id Taxonomy). For the British Isles (UK and Ireland) Blakemore (2004; 2007) lists 69 species, all thought to be introductions (Sims & Gerard, 1985; 1999:); *Polypheretima taprobanae* is a new record (*vide intra*) and *Pithemera bicincta* from recent literature, plus *Alma millsoni* (Beddard, 1891) from Kew Gardens that was overlooked by Sims & Gerard (1985, 1999) and others until now, which brings the current total to 72 species.

In France Bouché (1972) reported 180 taxa (not all valid species names) and these were commented on and added to by Zicsi and Csuzdi (1999) and Qiu and Bouché (1998) to give totals of about 300 names. Austria has ca. 60 lumbricids (48 and 60 listed by Zicsi (1981) plus *Lumbricus friendi*, *O. complanatus*, and *Criodrilus lacuum*). Edwards and Lofty (1977) gave these examples of European records: Czechoslovakia - 37 species (17 peregrine); Switzerland - 38 species (18 peregrine); Italy 57 - species (17 peregrine) [this total most recently updated as 86 megadriles from a total Oligochaeta of 304 species]; Germany - 24 species [19 peregrine, compared to 39 German lumbricidae listed online by Graff and Makeschin, (1980) and Bouch (1984)]. Bulgarian fauna is described by Zicsi and Csuzdi (1997) (54 spp); and Hungarian fauna is described by Csuzdi and Zicsi (2003). The Czech Republic species listed by Pizl (1994; 2002) described earthworms from the Czech Republic; this also has a faunal website (http://www.biolib.cz/en/taxon /id20098).

In Poland, Plisko (1965; 1973) surveyed the Polish lumbricids (32 spp.), and Perel (1977; updated by Blakemore, 2004; 2007) described those from the former USSR: (113 species: 107 Lumbricidae and 6 species from 5 other families and ca. 50% of these endemic). Slovenian and Balkan species are described by Mrsic (1985; 1991). Macedonia has ~29 species. Faunas of Greece, Portugal, Spain and Turkey (76 species) are found in Zicsi and Michalis (1981), Omodeo and Rota (2004), Trigo et al. (1990), Csuzdi et al. (2006; 2007), and Mısırlıoğlu et al. (2007).

Finland earthworms were catalogued by Edwards and Lofty (1977) and Terhivuo (1988; 1989). Swedish worms were described by Julin (1949);

Sweden also has a SWORM project database of some 20 earthworm species [http://www2.nrm.se/ev/dok/oligochl.html.en]. Danish taxa (fifteen Lumbricidae species in 8 genera) are described by Clausen (1987; 1993) and about 20 Lumbricidae plus a note about Megascolediae and Eudrilidae in hot-houses are listed on Lars Skipper's website [http://hjem.get2net.dk/lars_ skipper/regnorme3.htm, supported with details from Skov-og Naturstyrelsen database on introduced taxa www.skovognatur.dk/natur/ invasive_arter/ images/ introarter.xls 2007]. A few other North Eastern Atlantic Taxa (NEAT) are listed in Hansson (1998) [http://www.tmbl.gu.se/ libdb/taxon /neat_pdf/ NEAT*Annelida.pdf]. Volcanic Iceland has approximately 18 lumbricids, all introduced; and about 8 of the same species are known from glaciated Greenland (and Disko Island), as summarized by Omodeo (1957) and Blakemore (2006; 2007).

Few surveys are available from the Middle East, although El-Duweini and Ghabbour (1965) list some Egyptian worms and Pavlicek et al. (2003) have 34 taxa from the Mediterranean Lavant region tomas-pavlicek-biologie.net/ pages/ earthwormse.php; and 24 from Israel earthworms.elte.hu/Israel/ israel1.htm. Omodeo et al. (2003) list all known megadrile fauna of the Maghreb.

5.5. Earthworm Diversity in India[1]

India is located $8.4^{\circ} - 37.6^{\circ}$N latitude with an area of 3,287,797 km^2 and is a rich country as far as its biodiversity is concerned (Figure 5.1). Considering the past geological history of the Indian subcontinent (excluding Burma, Nepal and Pakistan) it has been divided into six well-defined physiographic regions depending upon topography, climate and vegetation; namely, the Western Himalayas, Eastern Himalayas, and Northeast ranges, Indo-Gangatic plains, Central high lands, Peninsular plateaus and Western ghats including Sahyadri and Nilgiri hills.

Different geographical divisions of the Indian subcontinent have been proposed from time to time.

[1] Source: Lecture delivered by Dr. J.M Julka in Training Workshop on *Inventorying Soil Biota* 17-23 October, 2005 at CSKHPKW. Palampur; YSPUHF, Nauni and Zoological Survey of India, Solan (HP), India.

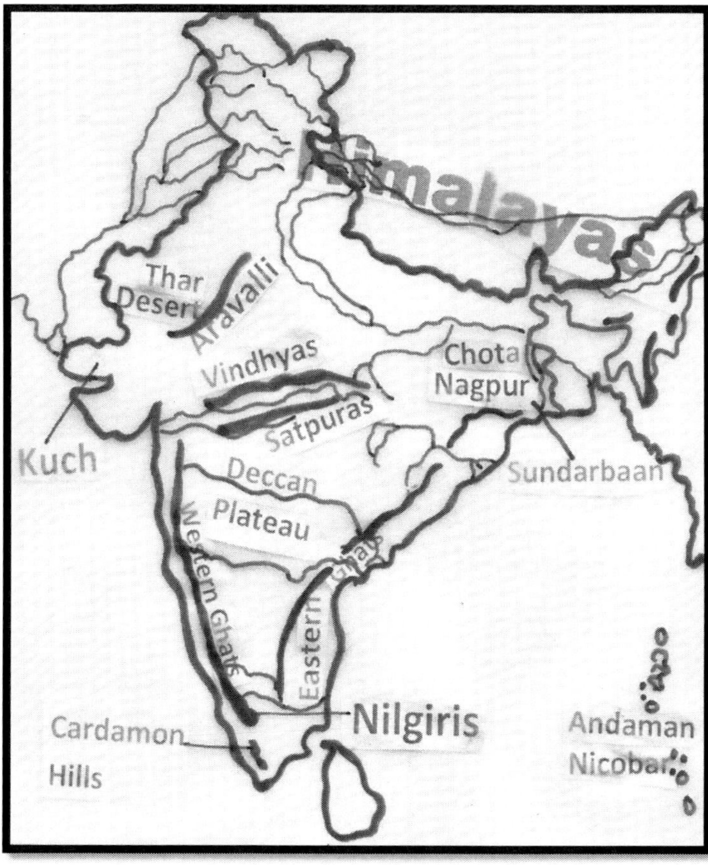

Figure 5.1. Physiographic and Zoo-geographic regions of India.

Smith (1931-43) divided this subcontinent into six geographical areas such as the desert area of the North West region (North-west frontier province; Punjab; Western Rajasthan as far as the Aravallis and Sind); Kashmir and Western Himalayas up to Nepal; the Indo-Gangatic plains and the Deccan (most of the Peninsular of India between 10°N and 20°N); mountains of the Malabar tract and Sri Lanka; the Chota Nagpur area. The Indo-Chinese sub region is represented by the Eastern Himalayas (from the western border to Nepal to the bend of the Brahmaputra whole of Burma except low lands in the South and the Northern part of Siam); Annam; Peninsula of Indo-China; Tenasserium and Peninsular South and the Northern part of Siam. About 3,600 species of earthworms have been identified globally (Reynolds, 1994) as shown in Figure 5.2.

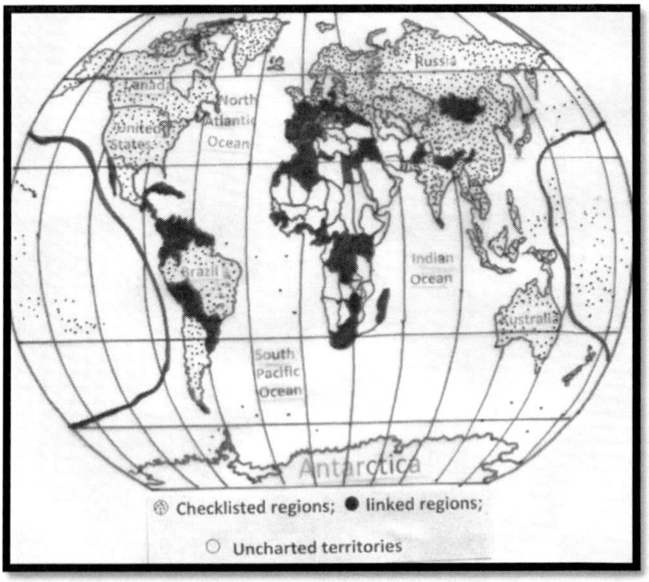

Figure 5.2. Regions of the globe with earthworm checklists.

India is a diverse country harboring a very high diversity of earthworms mostly concentrated in Western Ghat and Eastern Himalayas, both of which are recognized as biodiversity 'hot spots' (Figure 5.3).

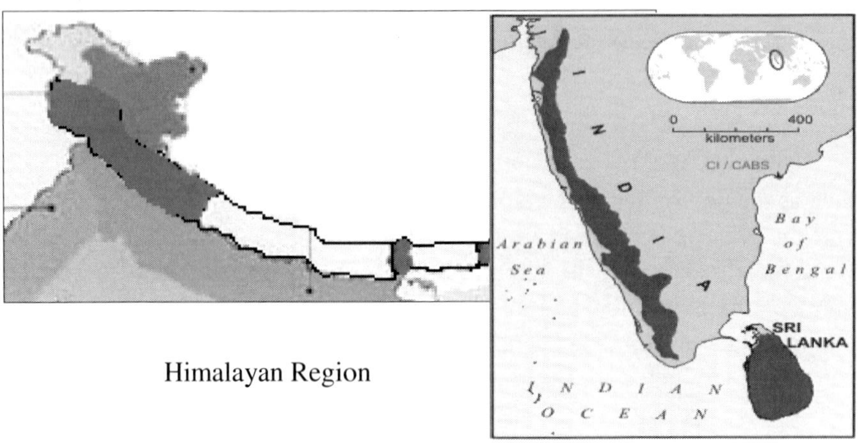

Himalayan Region

Western Ghats

Figure 5.3. Hot spots of biodiversity of Indian fauna.

Although the land area of India is only 2% of the total world's land mass, nevertheless, it supports about 10.5% of the total known global earthworm diversity (Figure 5.4). The Indian earthworm fauna is predominantly represented by native species which constitutes about 89% of total earthworm diversity in the country (Julka and Paliwal, 2005).

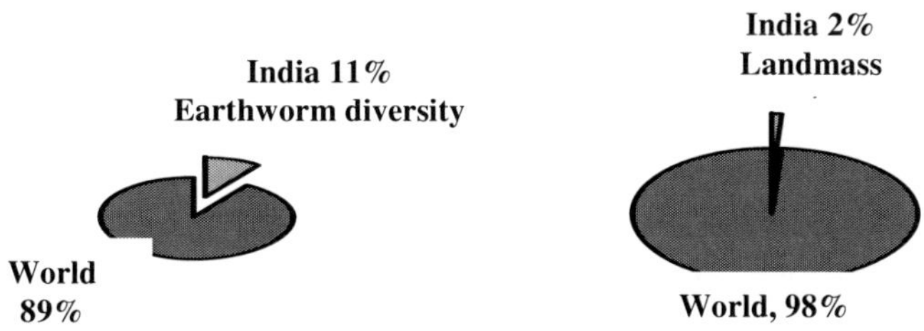

Figure 5.4. Comparison of Indian landmass and earthworm diversity with world.

The scientific explorations of earthworm diversity in India date back to the nineteenth century. The credit for naming the first earthworm species in the Indian sub continent went to Templeton (1844) when he discovered *Megascotex caerulens* from Sri Lanka. However, Perrier (1872) was the first to describe earthworm species from the Indian mainland (Figure 5.5). Subsequent noteworthy contributions on the taxonomy of Indian earthworms are those of Bouche (1971), Beddard (1902), Michaelsen (1903; 1910) and Stephenson (1914; 1916; 1917; 1922), significantly adding to the knowledge of earthworm diversity in Western Himalaya. Aiyer (1929), Gates (1937, 1938a, 1938b, 1939a, 1939b, 1942a, 1942b, 1945, 1947, 1972), Jamieson (1977), Julka (1976a, 1976b, 1978, 1981, 1983, 1988), Senapati et al. (1990), Julka and Paliwal (1993, 1994), Julka et al. (1997) and Paliwal and Julka (2005) have provided a comprehensive check-list for earthworms of Western Himalaya and Tamil Nadu. Stephenson (1923) listed 28 species of earthworms in the publication "Fauna of British India'' (volume on Oligochaeta), Western Ghats (Blakemore, 2006), West Bengal (Chaudhuri et al., 2008), North-east (Bhadauria and Saxena, 2010) and Gangetic plains (Verma et al., 2010) of India. The earthworm fauna of India is well reported compared to other Asian countries (excluding Myanmar).

Presently 418 valid species / subspecies of earthworms under 69 genera are known from the Indian Territory, including the islands of Andaman, Nicobar and Lakshadweep, and are placed in 9 families (according to Brinkhurst and Jamieson's classification of Oligochaeta and Gates' classification of Megascolecidae). Some exotic peregrine species of earthworms are also found and are now widespread in disturbed habitats following deforestation and intensive cultivation practices (Gates, 1982).

It is estimated that as many as 45 exotic species have been brought to India and mostly belong to union-territory from other Zoogeographical realms. Native species generally inhabit undisturbed natural ecosystems, but some of these are able to withstand disturbance and have acquired a widespread disturbance in different agro-ecosystems of the country. The most diverse families of earthworm species in India Megascolecidae comprise 136 species (119 native) of 13 genera; Octochaetidae with 134 species (129 native) of 30 genera and Moniligastridae with 79 species (all native) of 3 genera. Especially diverse earthworm genera of India belongs to Moniligastridae *Drawida* (68 species); the Megascolecidae *Perionyx* (46 species) and Megascolex (33 species); the Acanthodrilidae *Agrilopilus* (22 species) the Octochaetides *Eutyphoeus* (22 species), *Haplochaetella* (19 species) and *Octochaetona* (15 species). The high earthworm diversity in this country is primarily due to its geographical location with a wide latitudinal range (between 8.4' N and 37.6' N), complex topography, varied climate and past geographical history (linked to the ancient supercontinent of Gondwanaland from which it spread in the last Jurassic period and drifted to collide with the Asian Mainland to Eocene).

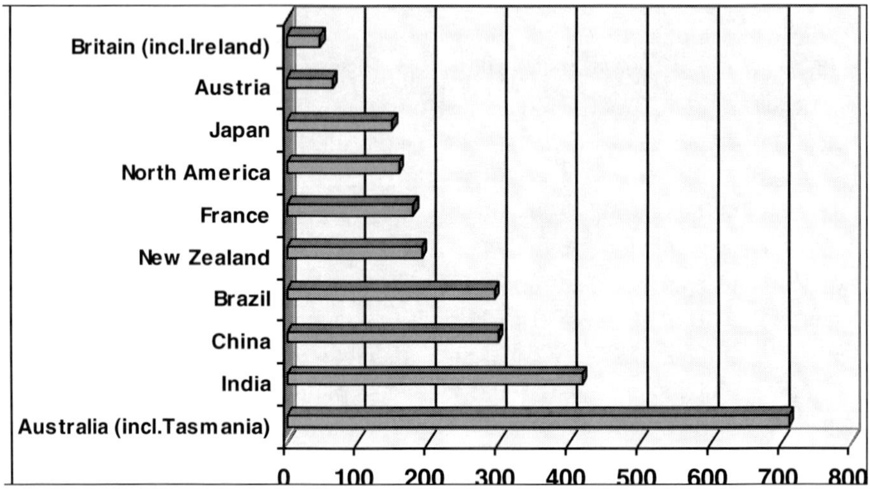

Figure 5.5. Global distribution of earthworm.

Earthworm biodiversity in India is predominantly represented by native species (357species), which constitute 88.8% of the total earthworm diversity of India. Although the majority of the Indian forms have a specific preference for natural habitats, a few are exotic and ubiquitous (Table 5.2). Indian species have successfully colonized different agro-ecosystems.

Table 5.2. Species density (Native / Exotic) of earthworms in different agro-climatic regions of India

S.No.	Agro-climatic Region	Species density		
		Native	Exotic	Total
1.	Western Himalaya	22	26	48
2.	Eastern Himalaya & NE Hills	85	25	110
3.	Lower Gangetic Plains	10	09	19
4.	Middle Gangetic Plains	24	05	29
5.	Upper Gangetic Plains	17	06	23
6.	Trans Gangetic Plains	04	02	06
7.	Eastern Plateau Hills	26	05	31
8.	Central Plateau Hills	27	12	39
9.	Western Plateau Hills	14	07	21
10.	Southern Plateau Hills	44	12	56
11.	Eastern Ghats and East Coast Plains	32	07	39
12.	Western Ghats and West Coast Plains	193	26	219
13.	Gujarat Plains & Hills	05	04	09
14.	Western Dry Region	03	01	04
15.	The Island	07	21	28

Julka et al. (2004) classified six earthworm diversity zones in India; namely, mega earthworm diversity, high earthworm diversity, medium earthworm diversity, low earthworm diversity, very low earthworm diversity and poor earthworm diversity zones.

5.5.1. Mega Earthworm Diversity Region

West coastal plains and Ghats are mega earthworm diversity regions which constitute 51.7% of known Indian species; among them 52% are native

Indian species which are endemic to this area and also comprise several native monotypic genera (Figure 5.6).

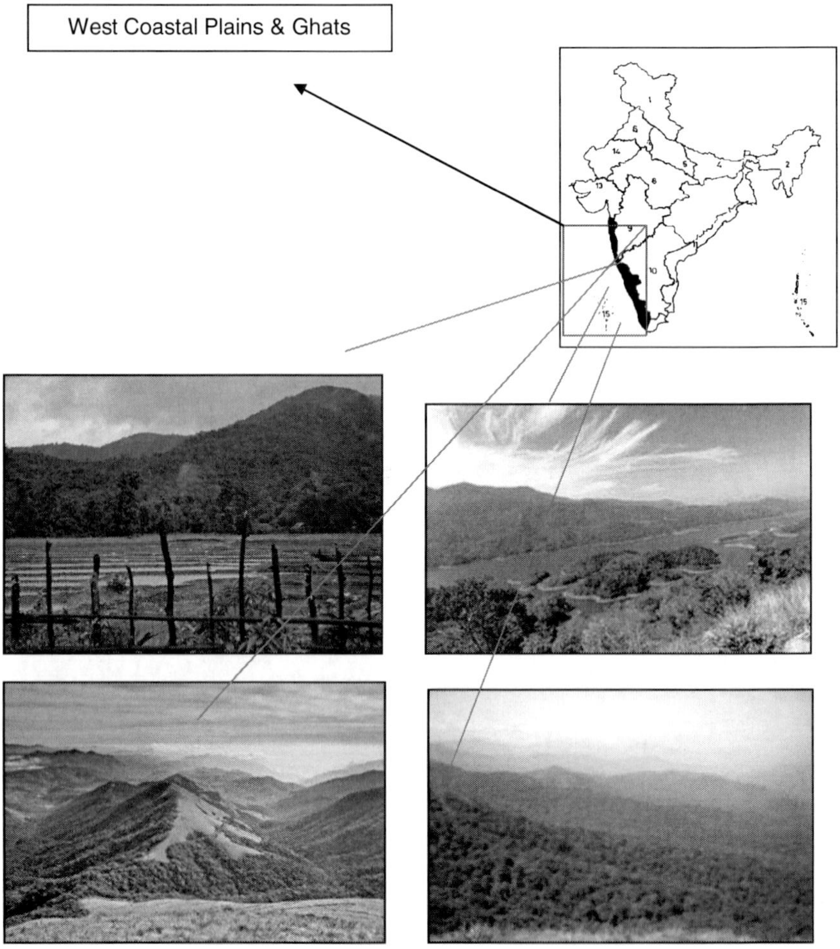

Figure 5.6. A view of mega earthworm diversity regions of India.

5.5.2. High Earthworm Diversity Region

Eastern Himalaya (Figure 5.7) is known as a high earthworm diversity region, which constitutes 26.2% of total known Indian species; among them 22% are native Indian species which are endemic to this area.

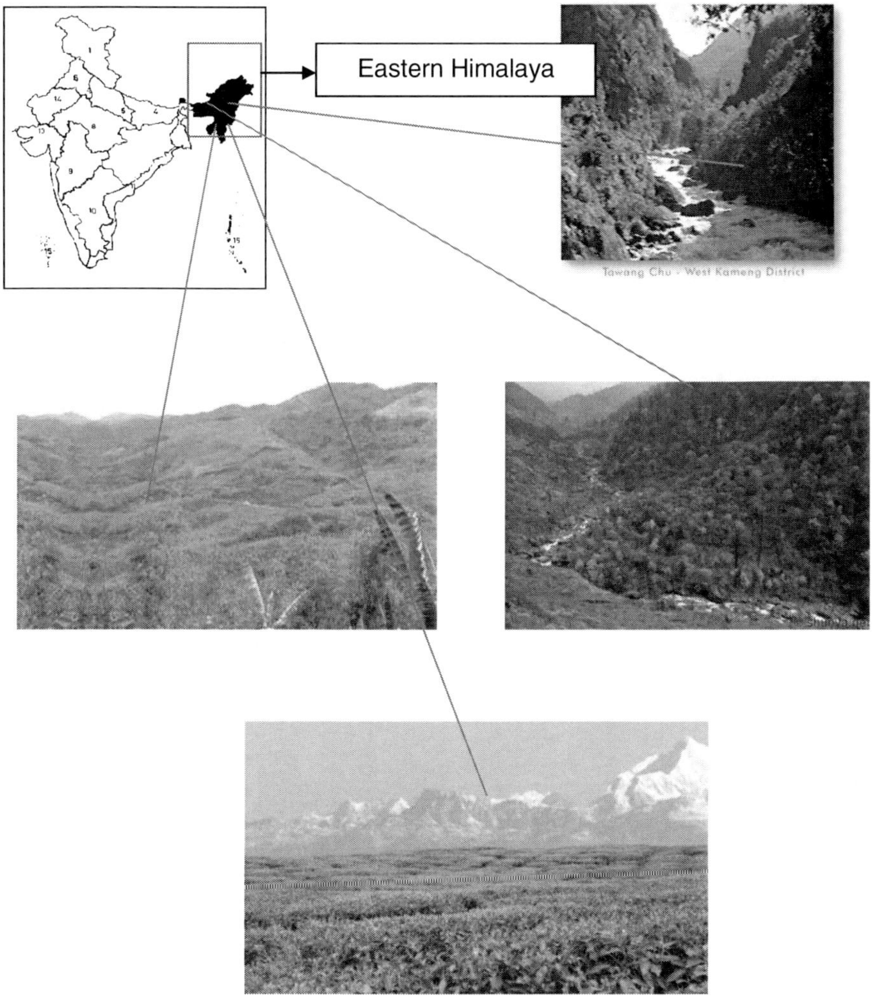

Figure 5.7. A view of high earthworm diversity regions of India.

5.5.3. Medium Earthworm Diversity Region

Western Himalayan and South Plateau hills are medium earthworm diversity regions of India. This comprises eleven percent of earthworms of total known Indian species; among them 2% are native species and endemic to this area.

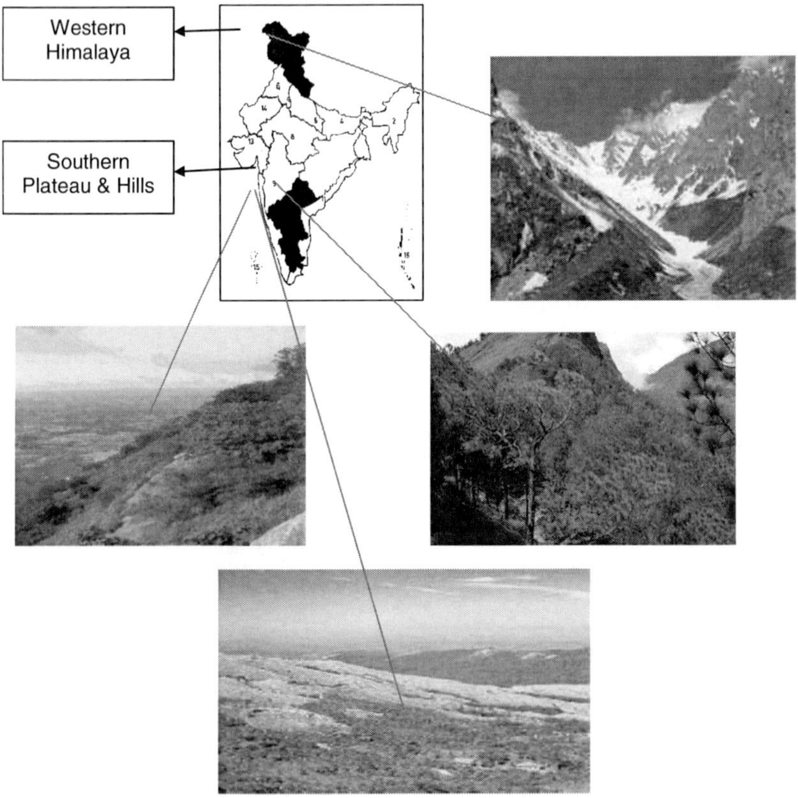

Figure 5.8. Medium earthworm diversity regions of India.

However, 13.4% species of total known Indian earthworms belong to Southern Plateau hills; among them ten percent are of native origin (Figure 5.8).

5.5.4. Low Earthworm Diversity Region

Eastern coastal plains and ghats, eastern plateau and hills, middle Gangetic plains, Central Plateau and hills are known as low earthworm diversity regions, which comprise 6.9-9.3% species; among them 4-8% species are endemic to this area (Figure 5.9).

Figure 5.9. Low earthworm diversity regions of India.

5.5.5. Very Low Earthworm Diversity Regions

Upper Gangetic plains, Lower Gangetic Plains, Western plateau and hills, The Islands hills are also known as very low earthworm diversity regions, which comprise 4.5 – 6.7% of total known Indian species; among them 2-3% are of native Indian species (Figure 5.10).

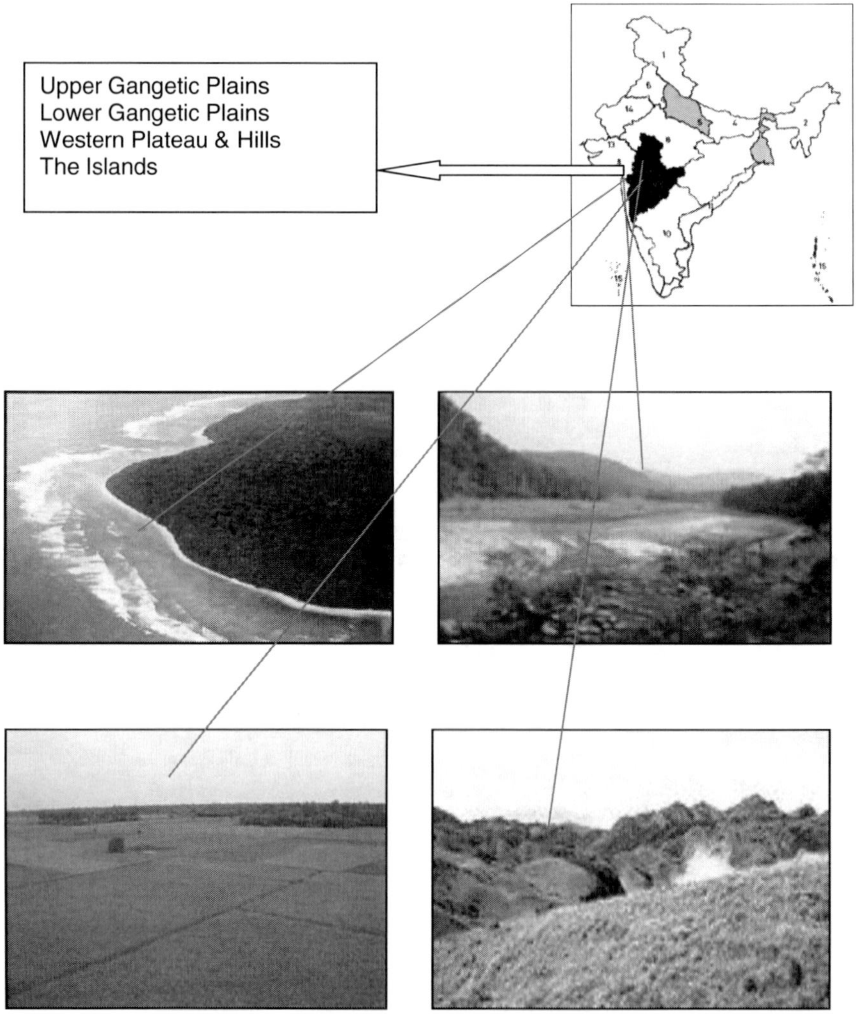

Figure 5.10. Very low earthworm diversity regions of India.

Hills are also known as very low earthworm diversity regions, which comprise 4.5 – 6.7% of total known Indian species; among them 2-3% are of native Indian species (Figure 5.10).

5.5.6. Poor Earthworm Diversity Regions

Trans Gangetic plains, Gujarat plains and hills and Western dry regions are known as poor earthworm diversity regions. Among them 0.9 – 3.8% of the total known are native Indian species and about 0-0.5% species are endemic to this area (Figure 5.11).

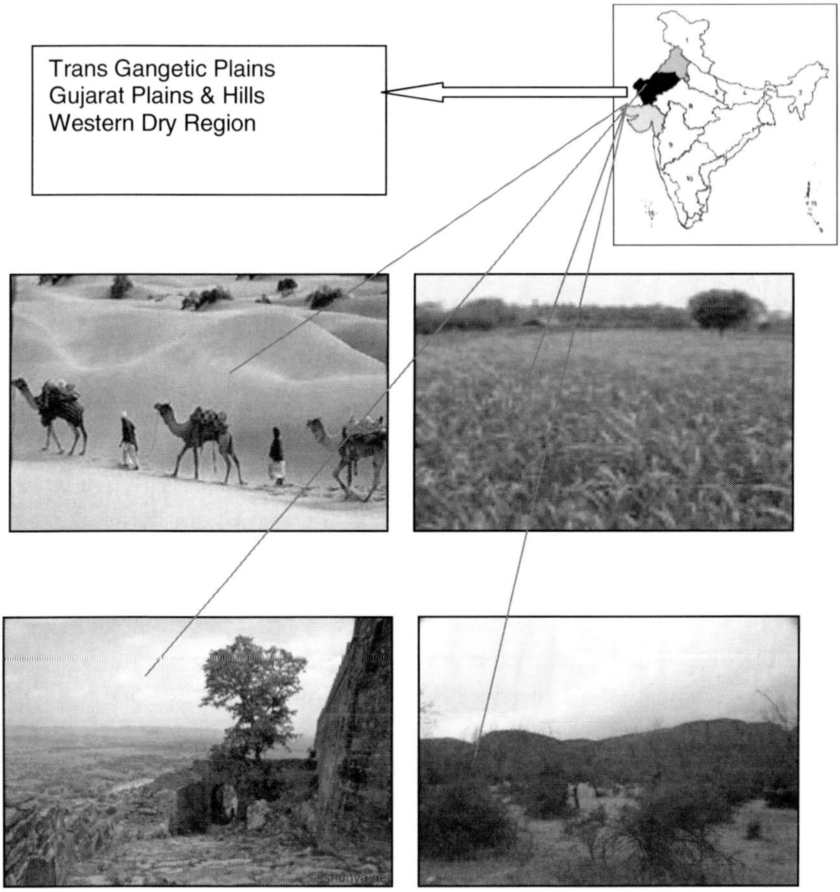

Figure 5.11. Poor earthworm diversity regions of India.

Verma et al. (2010) surveyed the Gangetic Plain of Uttar Pradesh, India, during August – October 2008, and collected 11 taxa of earthworms belonging to 6 Genera and 2 Families. This constitutes 26.3% of the total Indian earthworm fauna. Of these, 4 taxa are exotic with extra-Indian origin. A survey of earthworms in Doon Valley of the Western Himalayan region conducted by Verma and Shweta (2011) in September 2009-12 lists species belonging to 7 genera and 4 families. *Drawida nepalensis*, *Entyphoeus orientalis*, *Lampito mauritii* and *Perionyx sansibaricus* have been identified as potential species for vermicomposting with a preference for dung heap. Verma and Shweta (2011) surveyed earthworm bioresources with particular reference to native species of the northwest region of India and recorded a total of six species; namely, *Lampito mauritii* Mauritii, *Perionyx excavatus* Perrier, *Metaphire anomala* Michaelsen, *Drawida nepalensis* Michaelsen, *Ocnerodrilus occidentalis* Eisen and *Metaphire posthuma* vallant.

Apparent high diversity reports are mainly due to eco-taxonomy that records smaller species and recognizes the persistence of natives, combined with contributions from the pool of exotics originating from various temperate and tropical regions of the world.

5.6. Implications of Climate Change on Earthworm Distribution

Possible interactions between global change and earthworm invasions are of considerable interest (Bohlen et al., 2004; Frelich et al., 2006). First, the distribution of earthworms worldwide could be altered with changes in land use and climate. Land transformation provides more opportunities for invasion and, conversely, invasions enhance and drive land transformation (Hobbs, 2000). Deforestation and other human disturbances could create more habitats suitable for occupation by introduced species. Climate change has the potential to modify impacts of invasive species by affecting their sources, pathways and destinations (Sutherst, 2000). Furthermore, with a warming climate, the demarcation lines of earthworm distributions might advance poleward and to higher elevations, where few or no earthworms occur owing to continental and alpine glaciations (Hendrix et al., 2008). Some European Lumbricid species (*Dendrobaena octaedra* and *Eisenia nordenskioldi*) already occur near the Arctic Circle in Russia (Tiunov et al., 2006) and could be expected to advance more rapidly northward into thawing permafrost. Even at lower latitudes and altered temperature, adopted lumbricids in tropical areas are expanding to the range of regional peregrines, such as the pheretimoid species in East Asia. Warming at high elevations might be a particular problem for biodiversity conservation; high endemism in mountainous areas could be threatened by invading earthworms (Pop and Pop, 2006).

The impacts of earthworms on global change are likely to be complex; especially in areas previously devoid of earthworms, invasive earthworms may increase the decomposition of soil organic matter and release more CO_2 into the atmosphere (Potthoff et al., 2001; Speratti and Whalen, 2008) and in some cases also more N_2O (Rizhiya et al., 2007). In the longer term, earthworm activities may increase the sequestration of organic carbon in the soil *via* protection from decomposition within water-stable aggregates resulting from earthworm casting (Bossuyt et al., 2005; Martin, 1991).

5.7. General Body Plan of Earthworm

Earthworms are soft and fleshy as they do have neither an internal nor an external skeleton. They are a flexible, long bundle of muscles, especially designed for burrowing into soil and subterranean life. A worm's body is elongated and cylindrical and the surface of its skin is smooth and slimy. The underside (ventral surface) of the body is slightly flattened and lighter in color than the upper side (dorsal surface). It lacks a distinct head and external appendages, and tapers at both ends, although one is more pointed than the other. The sharper end is the anterior or head-end, whereas the blunter end is the posterior or tail-end.

The entire body is made up of several little rings, which are arranged symmetrically one beside each other and clearly demarcated by grooves or furrows. These rings are called segments. It is because of their ringed appearance that the earthworms are grouped under Phylum Annelida (derived from the Latin word *annulus* - 'a little ring'). 'Annelid' means ringed or segmented - a characteristic that separates these worms from other animal groups.

The general body plan is tube-like. The outer tube is the body wall and the inner tube is the digestive tract or gut, which runs throughout the length of the body. The space between the two tubes, commonly called the body cavity, is filled with coelomic fluid which acts as a hydrostatic skeleton. Each segment is a separate fluid-filled compartment surrounding a digestive tract. The compartments of the segments are separated from each other by thin partitions (known as septa) that are attached to the body wall at sites of inter-segmental furrows. Many of the worm's internal organs are also segmented, occurring as separate units in each segment.

5.8. Body Size of Earthworms

Earthworms vary in size considerably. The smallest earthworms are about 10 mm long. In India these are represented by some exotic species like *Microscolex phosphoreus* and *Bimastos parvus*. The largest earthworms (*Microchaetus rappi*) in the world are found in South Africa and are reported to be about 7 meters long. One of the large-sized earthworms in Australia, the Giant Gipsland earthworm (*Megascolex australis*), measures over 2 meters in length. There are giant-sized earthworms in India too. The largest Indian earthworm is *Drawida nilamburensis*, which is around one meter long (Table 5.3) and lives deep in the soil on the banks of river Cheliyar in Kerala.

Table 5.3. List of Giant Indian earthworms*

Species Family	Length (mm)	Distribution
Family: Moniligastridae		
Drawida grandis	520	Nilgiris (Tamil Nadu)
Drawida naduvatamensis	500	Nilgiris (Tamil Nadu)
Drawida nilamburensis	1000	Nilgiris (Malbar, Kerala)
Family: Megascoecidae		
Megascolex imperatrix	650	Nilgiris (Tamil Nadu)
Megascolex konkanensis	415	South India
Megascolex konkanensis longus	570	Annamalai Hills
Perionyx macintoshi	375	Eastern Himalaya
Family: Octochaetidae		
Periodochaeta pellucida	450	Nilgiris (Tamil Nadu)
Eutyphoeus gammiei	405	Eastern Himalaya & Northeast Hills

* Source: Julka, J.M., 2008. Know your earthworms. Foundation for Life Sciences & Business Management, Solan, India.

5.9. Body Structure

The first segment of the earthworm's body is called the peristomium, which bears a small crescent-shaped mouth at its terminal end. A small tongue-like lobe of skin, the prostomium, projects out in front of the first body segment. It is located just above the mouth. It helps the earthworm's sense of light and vibrations and facilitates its way through the soil. The worm's skin cells are also sensitive to light, and can sense touch and chemicals. The last segment of the body is called the periproct and contains the anus. Earthworms cannot see, hear or smell.

A mature earthworm shows a prominent circular band-like swelling in the skin near to the anterior end. It is called the clitellum and covers over a few cocoons at breeding time. It can be flared, non-flared, saddle-shaped or annular. It may be white, orange-red or reddish-brown in color. The number of segments to where clitellum begins and the number of segments that make up the cliellum are important in identifying earthworms. Certain areas of skin on the underside of mature earthworms may be somewhat depressed or swollen in the form of genital markings or tubercula pubertalis. Earthworms are hermaphrodites; each worm possesses both male and female reproductive parts with the opening (also known as pores) to the exterior on its underside.

5.10. Earthworms' Habitat

Earthworms are present in almost every type of soil but the healthier the soil, the greater the numbers. They are generally found in the top 30 to 45 cm of soil. As they breathe through their skin, they must have a moist environment to assist them in respiration. Earthworms would either burrow deeper (up to 2 meters) or migrate to another location of soil if it becomes exclusively dry. They avoid coarse (sandy) and very acidic soils. Certain kinds of earthworms are found in large numbers in dead organic materials like compost, manure, all sorts of dung, leaf litter, humus etc. and in kitchen drainage and soils wet with effluents. Some earthworms also live in sub-aquatic conditions near freshwater aquatic bodies and in paddy fields, and one particular species, *Pontodailus litoralis*, is able to thrive in salt water marshes near seas. A few species can live under snow on high mountains.

5.11. General Habits of Earthworm

Earthworms are basically nocturnal and to avoid light, they live below ground or under leaf litter. They spend most of their time burrowing in the soil. The worms construct burrows by two methods: In loose soil, they press their setae against the wall of the burrow to hold their position and at the same time push aside soil particles with their protruding lobe-like prostomium. They can also literally eat their way through the soil. Since earthworms need moisture for their survival, they will gradually burrow deeper if the topsoil becomes dry. They also move deeper during the hot and winter months. They cannot regulate their body temperature. However, not all earthworms make burrows in the soil. Usually, worms living in leaf litter or dung avoid this activity.

At night, especially when the soil is moist and the atmospheric temperature is lower, earthworms come out of their burrows for feeding and sometimes for mating. During feeding, the worm protrudes its head end out of the burrow to grab and pull a small piece of dead organic matter by its mouth. If disturbed, it can quickly withdraw into the burrow at dawn in response to first light, which is sensed by its light-sensitive head end.

Sometimes, after a heavy rainfall, earthworms are seen in large numbers crawling on the ground. There are four main presumptions for this behavior. Firstly, the waterlogged soil has insufficient oxygen for the worms. Therefore, earthworms come to the surface to get the oxygen they need and breathe more easily, although they have been known to survive underwater for several hours if there is oxygen in it. Secondly, earthworms emerge from their burrows to mate, but this behavior is limited to some species. Thirdly, the worms come out of the soil for colonizing new areas using high moisture conditions on the soil surface, but this activity during the daytime exposes them to the dangers of lethal ultra-violet rays of sunlight and various kinds of predators, besides being trampled under the feet of animals. Fourthly, there are several kinds of other below-ground animals as well and the carbon dioxide produced by them during respiration dissolves in rain water to form carbonic acid, which makes the soil highly acidic for the worms to live in.

The soil swallowed by earthworms for making burrows passes through its body and is left over on the soil surface as fecal deposits, which are known as worm castings. Some species deposit their castings in their burrows. The surface worm casting may take different forms, depending upon the species. There may be little heaps of spherical or oval pellets, small to large towers of coiled tubes and small mounds of threads or beaded strings. The castings in some Indian earthworms are known to produce from 14 to 778 tons of casting/ha/annum in India, compared to the larger amount of 2100-2600 tons/ha/annum by African worms.

5.12. Food and Feeding Behavior of Earthworms

Earthworms are omnivorous and consume organic matter that consists of rotting plant material and decomposing remains of other animals. They do not feed on healthy plants. Earthworms also eat other living organisms, such as nematodes and microorganisms like protozoans, bacteria and fungi. A group of earthworms (*Agastrodrilus*) from the Ivory Coast in Africa is carnivorous, eating other species of earthworms.

Recent studies show that it is actually microorganisms growing on rotting materials that provide the main source of nutrition, rather than the decaying material itself. Thus, anything that will support microbes is potentially suitable for the earthworms. That is why it is recommended first to allow the rotting of the organic wastes (for developing a flourishing microbial community) and then to add worms to it. Along with the organic material predigested by microorganisms, earthworms also ingests small particles to help in fragmenting ingested organic matter into smaller particles of sand and soil, which collect in their gizzard. The coarse particles help in fragmenting ingested organic matter into smaller pieces for easy digestion in the gut of an earthworm.

According to their feeding habits, earthworms are generally divided into three main groups. The first group feeds mostly on decaying litter or animal dung; the second group consumes the bulk of poor subsoil which has very little organic matter; and the third group eats a large amount of mud along with decayed debris accumulated on the margins or the bottom of water bodies. Depending upon the species, in just one day earthworms can consume food up to their own body weight.

5.13. Defense in Earthworms

Earthworms defend themselves in a variety of ways. Those living in tunnels escape danger by quickly withdrawing into their burrows. Those living near or on the surface have the ability to move quickly, sometimes in a serpentine manner (for example, *Amynthas cortices*). Some of them - like *Perionyx excavatus* and *Ramiella pallada* - can jump, thrash wildly about and produce a series of rapid lashing movements. A few species (*Hoplochaetella inornata and H. stuarti*) will even purposely break off a tail segment when threatened by a predator. Some, like *Metaphire posthuma*, will twist and coil their bodies, or produce bad-smelling fluids like the *Eisenia fetida*. *Amynthas alexandri* and *Hoplochaetella suctoria*, when touched, become limp as if they are dead. However, species living deep in the soil have few defensive behaviors.

Earthworms have no eyes, but they do possess cells which are sensitive to light. They are photophobic to some degree, meaning they react negatively to bright light. The light is detected through 'sensors' located in the skin, mostly near the head and tail ends. They don't have ears either, but they can feel vibrations in the ground. However, worms have a very good sense of smell, which they use to locate food, water and even their mates.

Many species of earthworms suspend their activities and undergo dormancy (diapauses) when the soil becomes too dry or the soil temperature is either very low or very high. During this stage, a worm stops feeding, empties its gut and constructs a small spherical chamber in the soil, which is lined with a thick coat of sticky fluid. The worm rolls itself into a tight ball or loose knot in the chamber to tide itself over adverse environmental factors. In these conditions, it loses weight but without any tissue dehydration. As and when conditions become favorable, the worm emerges from dormancy to resume activities like burrowing, feeding, reproduction etc.

5.14. Reproduction in Earthworms

An adult earthworm is easily recognized by the presence of a swollen whitish or pinkish band (clitellum) at a little distance behind its head end. Earthworms are hermaphrodites (meaning that each worm has both male and female sex organs). The hermaphroditic way of life in earthworms is considered to be an adaptation to secondary life, which provides few contacts between individuals; the main advantage of hermaphroditism is the assurance of a reproductive strategy that involves mate recognition, mutual exchange of sperm and release of eggs within a protective and nutritive capsule (cocoon) produced by each mating worm.

In India, earthworms mostly breed during the monsoons. The breeding is, however, continuous in vermicomposting species in culture bins, provided the conditions are favorable. Sexually matured worms secrete a fluid with a scent, which attracts potential mates to them. By keeping a high population density in a cutting bin, the chance of two suitable worms finding each other is greatly increased, which is why cultured worms breed far more prolifically than their counterparts in the wild.

During mating, two worms lie side by side with their heads pointed in opposite directions, making close contact along anterior parts of their bodies. They align in such a way that the clitellum of one worm lies opposite to the segments that bear openings of sperm spring sacs (spermathecae) of the other. With the underside held firmly together by tiny bristles (setae), they excrete a sticky fluid that coats both worms and binds them together by tiny bristles (setae), preventing them from being easily pulled apart.

There are generally two ways of exchanging sperm between two mating worms, depending upon the location of their male pores in relation to their clitella. At the time of mating, the male pores in these worms are just in line with the spermathecal pores of each other, and the sperm from one worm is directly deposited in to the spermathecae of the other. Depending upon the species, the worms remain in the mating position for a period of time ranging from several minutes to a couple of hours. After the exchange of sperm, they separate and move away in their different ways to complete the reproductive process, each carrying the sperm from the other in its spermathecae.

After the worms have separated, the clitellum of each worm secretes a jelly-like substance called albumin. This material hardens on the outside but remains sticky underneath, and forms a band-like structure around the worm. The worm wriggles out of the hardened mucus band which closes at both ends, forming a cocoon with the sperm and eggs inside. Fertilization takes place inside this cocoon. The worm will continue to produce cocoons until the sperms received from its mate are completely used up.

Depending upon the species, earthworms produce a varying number of cocoons (Table 5.4). Under normal conditions a worm will make around 50 cocoons a year. Earthworms also start producing more cocoons under adverse conditions like the depletion of the food source, low temperatures and drying of soil (bedding in culture bins). It is a reproductive strategy to ensure the continuation of future generations. In order to get increased cocoon production, some vermiculturists actually stop watering culture bins to dry out vermibeds.

Under unfavorable conditions such as dryness, cocoons can remain inactive for years (sometime as long as 30-40 years) and hatch when conditions become favorable. Cocoons are also very tough and can tolerate a lot of handling and rough treatment. The shape of the cocoon differs very much among earthworms. It may be spindle-shaped (*Perionyx excavatus*), egg-shaped (*Lampito mauritii*), spherical (*Pontoscolex corethrurus*), lemon-like (*Eisenia fetida*), tapered lemon-shaped (*Eudrilus eugeniae*) or onion-shaped (*Drawida nepalensis*).

Table 5.4. Young worms/cocoon and their incubation period in some Indian earthworms

Species	Average number of young worms/cocoons	Incubation period (days)
Dichogaster bolani	2	7
Perionyx excavatus	1	13-14
Eudrilus eugeniae	2	15-30
Eisenia fetida	2-3	14-230
Lampito mauritii	1-3	15
Drawida nepalensis	1-2	25
Pontoscolex orethrurus	1-2	29
Polypheretima elongate	1-2	50
Eutyphoeus gammiei	1	110

A newly-laid cocoon is usually creamy white, and then turns yellowish on the second or third day and gradually undergoes various shades from brown to dark brown or dark purple by the time it is ready to hatch. Although the number of fertilized ova in each cocoon generally ranges from one to twenty, often only one or two survive and hatch.

The development or incubation time of a cocoon varies with the species and environmental factors. It varies from 7 to 55 days in worms living in warmer areas and 30 to 210 days in colder regions. Worms hatched from cocoons in a given material will tend to be much better adopted than any adult worms introduced to the same material. The worms that are raised in manure if fed on food scraps would initially roam away. Worms continue to grow once they reach sexual maturity, but once at this stage, there is a much slower increase in weight until the disappearance of the clitellum indicates the onset of old age or senescence. During this period there is a slow decline in weight until the death of the worm. The lifespan of an earthworm varies in different species. Under field conditions, most surface-dwelling worms live less than a year as they are easily exposed to several kinds of predators, as well as very hot and dry conditions. It is estimated that certain species can live up to about 3 years. However, individuals of some species have been kept alive for longer periods in laboratory cultures. For example, *Eisenia fetida* has been kept alive for 4 years, *Lumbricus terrestris* for 6 years and *Allalobophora longa* for 14 years. There are no hybrid earthworms, as they do not mate between the species. This is because (besides any genetic incompatibilities) the sex organs of different species do not correspond.

5.15. Ecological Groups of Earthworms

Earthworms are grouped into three broad ecological groups based on their habitat, feeding and burrowing habits:

A) Litter and surface-dwelling species (*Epigeic* – 'upon the earth')
B) Soil-dwelling or upper soil surface (*Endogeic* – 'within the earth')
C) Deep-burrowing species (*Anecic* – 'out of earth')

Earthworms

Epigeic

- Litter feeder
- Litter dweller
- Pigmented
- No burrows
- Small size

Endogeic

- Rich soil feeder
- Top soil dweller
- No pigmentation
- Horizontal burrows
- Small size

Anecic

- Litter + soil feeder
- Soil dweller
- Dorsally pigmented
- Extensive vertical burrows
- Large size

5.16. Earthworm Species Used in Vermiculture

As many as 418 species and subspecies of earthworms are known from India, and these represent both exotic and native species. Of these, about a dozen have been identified as useful in vermicomposting systems. However, in India, two exotic species - namely, Eudrilus eugeniae and Eisenia fetida - and one native species, *Perionyx excavatus*, are extensively used in the vermicomposting industry.

These species have been evaluated based on their:

- Tolerance to a wide range of environmental conditions and fluctuations
- Tolerance to disturbance
- Rapid growth
- High breeding rate

5.17. Some Significant Properties of Earthworms

(i) *Sensitive to light, cold and dryness:* Earthworms are very sensitive to light, cold and dryness. They tend to migrate temporarily into deeper layers of soil when subjected to conditions that are light, too cold or too hot. This is of great importance to their survival, especially in cold winters and hot summers.

(ii) *Adapted to survive in a harsh environment:* Some species, e.g. *Eisenia fetida*, are highly adapted to survive in 'harsh' conditions where no creature on earth can survive. After the Seveso chemical plant explosion in 1976 in Italy, when a vast inhabited area was contaminated with certain chemicals including the extremely toxic TCDD (2,3,7, 8-tetrachlorodibenzo-*p*-dioxin), several fauna perished but for the earthworms that alone were able to survive. Earthworms which ingested TCDD contaminated soils were shown to bio-accumulate dioxin in their tissues and concentrate it on average 14.5-fold (Satchell, 1983). *E. fetida* was used as the test organisms for different soil contaminants and several reports indicated that *E. fetida* tolerated 1.5% crude oil (containing several toxic organic pollutants) and survived in this environment. Earthworms also tolerate high concentrations of heavy metals in the environment.

(iii) *Ability to degrade most organic wastes rapidly into nutritive vermicompost:* Research into vermiculture has revealed that worms can feed upon a wide variety of organic wastes and provide a sustainable solution for total waste management. Farm wastes, animal wastes, garden wastes from homes and parks, sewage sludge from the municipal wastewater and water treatment plants, and wastewater sludge from paper pulp and cardboard industry, brewery and distillery, sericulture industry, vegetable oil factory, potato and corn chips manufacturing industry, sugarcane industry, guar gum industry, aromatic oil extraction industry, and the logging and carpentry industry offers excellent feed material for vermicomposting by earthworms.

Even 'flyash' (rich in nitrogen) from the coal power plants once considered a 'biohazard' can be composted by earthworms and converted into organic fertilizer. The worms digest the waste and convert a good part of it into mineral-rich nutritive vermicompost which is much superior to all the conventional composts.

Livestock rearing waste such as cattle dung, pig and chicken excreta makes excellent feedstock for earthworms. Animal excreta containing excessive nitrogen components may require mixing with carbon-rich bulking agents (straw, sawdust, dried leaves and grasses, shredded paper waste, etc.) to maintain proper C/N ratio. Paunch waste materials (gut contents of slaughtered ruminants) from abattoirs also make good feedstock for earthworms.

The worms secrete enzymes proteases, lipases, amylases, cellulases and chitinases in their gizzard and intestine, which bring about rapid biochemical conversion of the cellulosic and the proteinaceous materials in the waste organics. Earthworms convert cellulose into its food value faster than proteins and other carbohydrates. They ingest the cellulose, pass it through their intestine, adjust the pH of the digested (degraded) materials, cull the unwanted microorganisms and then deposit the processed cellulosic materials mixed with minerals and microbes as aggregates called 'vermicasts' in the soil.

Most earthworms consume, at best, half their body weight of organics in the waste in a day. *Eisenia fetida* is reported to consume organic matter at a rate equal to its body weight every day. Earthworm participation enhances the natural biodegradation and decomposition of organic waste from 60 to 80%. Studies indicate that at optimum conditions of temperature (20-30°C) and moisture (60-70%), about 5 kg of worms (numbering approx. 10,000) can vermiprocess 10 ton of waste into vermicompost in just 30 days (Kumar et al., 2010). Upon vermicomposting, the volume of solid waste is significantly reduced from approximately 1 cum to 0.5 cum of vermicompost.

(iv) *Reinforce decomposer microbes to promote rapid waste degradation:* Earthworms promote the growth of 'beneficial decomposer aerobic bacteria' in waste biomass in several ways: by improving 'aeration' through burrowing actions; by releasing 'chemical mediators' along their gut and body surface and indirectly through protozoa which they activate, which act at low concentrations on microbial metabolism, as vitamins or as chemical catalysts. Earthworms host millions of decomposer (biodegrader) microbes in their gut (as they devour them) and excrete them in soil along with nutrients nitrogen (N) and phosphorus (P) in their excreta. The nutrients N and P are further used by the microbes for multiplication and vigorous action. Edward and Fletcher (1988) showed that the number of bacteria and 'actinomycetes' contained in the ingested material increased up to 1000-fold while passing through the gut. A population of worms numbering about 15,000 will in turn foster a microbial population of billions of millions. Singleton et al. (2003) studied the bacterial flora associated with the intestine and vermicasts of the earthworms and found species like *Pseudomonas, Mucor, Paenibacillus, Azoarcus, Burkholderia, Spiroplasm, Acaligenes and Acidobacterium*, which has the potential to degrade several categories of organics. Acaligenes can even degrade PCBs and *Mucor* can degrade dieldrin.

Under favorable conditions, earthworms and microorganisms act symbiotically and synergistically to accelerate and enhance the decomposition of the organic matter in the waste. It is the microorganisms which break down the cellulose in the food waste, grass clippings and the leaves from garden waste.

(v) *Ability to kill pathogens and disinfect its surroundings:* The earthworms release coelomic fluids that have anti-bacterial properties and destroy all pathogens in the medium which they inhabit. They also selectively devour the protozoa, bacteria and fungus as food. They seem to realize instinctively that anaerobic bacteria and fungi are undesirable (causing rotting and foul odor) and so feed upon them preferentially. They also produce 'antibiotics' and kill the pathogenic organisms in their surroundings. This attribute of earthworms is very useful in the composting of waste whereby the end-product becomes 'disinfected', 'odorless' and free of harmful microbes. The removal of pathogens, fecal coliforms *(E. coli), Salmonella* sp., enteric viruses and helminth ova from human waste appears to be much more rapid when they are processed by *E. fetida*. Above all, *E.coli* and *Salmonella* are greatly reduced (Kumar et al., 2010).

(vi) *Ability to bio-accumulate toxic chemicals and detoxify the medium in which it lives:* Several studies have found that earthworms effectively bio-accumulate or biodegrade several organic and inorganic chemicals including 'heavy metals', 'organochlorine pesticide' and the lipophilic organic micropollutants like 'polycyclic aromatic hydrocarbons' (PAHs) residues in the medium which they inhabit. No farmlands in the world today where heavy use of agrochemicals was made in the wake of the green revolution are free of organic pesticides. Several studies have found a definite relationship between 'organochlorine pesticide' residues in the soil and their amount in earthworms, with an average concentration factor (in earthworm tissues) of about 9 for all compounds and doses tested. The capacity for heavy metal removal by earthworms is of particular significance when using vermicomposts made from urban solid wastes. Urban waste may contain considerable heavy metals and when processed by earthworms only can they become free of heavy metals.

(vii) *Ability to tolerate and reduce soil salinity:* Studies indicate that *Eisenia fetida* can tolerate soils nearly half as salty as seawater (i.e. 15 gm/kg of soil) and also improve their biology and chemistry (average seawater salinity is around 35 g/L). Farmers at Phaltan in the Satara district of Maharashtra, India, applied live earthworms to their sugarcane crop grown on saline soils irrigated by saline ground water. The yield was 125 tons/hectare of sugarcane and there was a marked improvement in soil chemistry. Within a year there was 37% more nitrogen, 66% more phosphates and 10% more potash. The chloride content was reduced by 46%. Farmers in the Sangli district of Maharasthra, India, grew grapes on eroded wastelands and applied vermicasting at 5 tons/hectare. The grape harvest was normal with an improvement in the quality, taste and shelf life. Soil analysis showed that within one year pH was reduced from 8.3 to 6.9 and the value of potash increased from 62.5 kg/ha to 800 kg/ha. There was also a marked improvement in the nutritional quality of the grape fruits.

(viii) *Worms improve the total physical, chemical and biological quality of soil:* Earthworms are found in a wide range of soils representing 60-80% of the total soil biomass. Significantly, t he worms lead to total improvement in the quality of soil and land where they inhabit and also enhance total plant growth and crop productivity. One acre of fertile land may contain more than 50, 000 earthworms of diverse species. They play a major role in 'renewing soil fertility' by continuously burrowing, ingesting, turning, mixing, aerating and improving drainage of the soil and are regarded as a 'biological indicator' of soil fertility. They have even been introduced into reclaimed soils successfully to restore their fertility. Earthworm activity is so prolific that, on average, 12 tons/ha/year soil or organic matter is ingested by this population; leading to the upturning of 18 tons of soil/year the world over at this rate may mean a 2 inch humus layer over the globe. Earthworms can contribute between 20 to 40 kg nitrogen/ha/year in soil, in addition to other mineral nutrients and plant growth regulators and an increase in soil fertility and plant growth by 30-200%.

Earthworms, when present in soil, inevitably work as 'soil conditioners' to improve its physical, chemical and biological properties and also its nutritive value for healthy plant growth. This they do by soil fragmentation and breakdown of organic matter in soil and the release of nutrients, and secretion of plant growth hormones, proliferation of nitrogen-fixing bacteria, increasing biological resistance in crop plants. All these worm activities contribute to improved crop productivity. Worms swallow a large amount of soil with organics (microbes, plant and animal debris) every day, grinding them in their gizzard and digesting them in their intestine with the aid of enzymes. Only 5-10 percent of the chemically digested and ingested material is absorbed into the body and the rest is excreted out in the form of fine, mucus-coated granular aggregates called 'vermicastings' which are rich in NPK (nitrates, phosphates and potash), micronutrients and beneficial soil microbes. The organic matter in the soil undergoes 'humification' in the worm intestine in which the large organic particles are converted into a complex amorphous colloid containing 'phenolic' materials. About one-fourth of the organic matter is converted into humus. The humic acid has a very good impact on plant growth. The colloidal humus acts as a 'slow release fertilizer' in the soil.

(viii) *Worms provide high levels of bio-available nutrients in balanced form for plants*: Earthworms' excretion (vermicastings) in the soil carry ammonia, nitrates, nitrogen, phosphorus, magnesium and other micronutrients and nitrogen-fixing microbes. Earthworms mix organic and inorganic, living and nonliving elements indiscriminately and smear the milieu with mucus, urine and feces to form balanced plant nutrient. They produce extra soil nutrients from grinding rock particles and by enhancing atmospheric nitrogen fixation. They mineralize the nitrogen (N) and phosphorus (P) in the waste to make it bio-available to plants as nutrients. They ingest nitrogen from the waste and excrete in the mineral form as ammonium and muco-proteins. The nitrogenous waste excreted by the nephridia of the worms is plant-available as it is mostly urea and ammonia. The ammonium in the soil is bio-transformed into nitrates. The more significant is 'organic nitrogen' that does not accumulate in food products in a concentration that accumulates in food grown on chemical nitrogen (urea) posing a health risk.

(ix) *Nitrogen (N) contribution to soil:* Barley and Jennings (1959) reported that worms significantly contribute nitrogen (N) contents to soil by over 85%. When the young growing worms were fed with a soil containing finely ground leaf litter (containing nitrogen in non-bioavailable forms for plants), about 6% of the ingested nitrogen was excreted in bio-available forms for the plants. After 28 weeks soil with living worms contained 75 ppm of nitrate nitrogen, compared with the control soil which contained 45 ppm. Patil (1993) found that earthworms recycle nitrogen in the soil in a very short time and the quantity of nitrogen recycled is significant, ranging from 20 to 200 kg N/ha/year. Worms increase nitrogen levels in the soil by adding their metabolic and excretory products (vermicast), mucus, body fluid, enzymes and decaying tissues of dead worms. They also contribute nitrogen indirectly through the fragmentation of organic materials and grazing on soil microorganisms.

Earthworms' tissue contains about 10% N on a dry weight basis. Whalen et al. (1999) reported that living worms release nitrogen from their bodies and after death it is rapidly decomposed, releasing all nitrogen into the soil. Christensen found that 50% of the N in dead worm tissues was mineralized in 7 days, while Satchell (1983) found it to be 70% in 10-20 days and the N was converted to NO_3-N which is the bio-available form on nitrogen to crop roots. The release of mineral N after the death of the earthworms could be significant since worm biomass can turn over up to 3 times a year in farm soil. The study estimated the direct flux of nitrogen through earthworm biomass in farm soils (agro-ecosystems) ranging from 10-74 kg N/ha/year.

(x) *Phosphorus (P) contribution to soil:* It is well established that worm casts are richer in inorganic phosphorus compounds extractable in water than the surface soil ingested. Graff (1981) and Sharpley and Syers (1976) found that exchangeable phosphorus (P) measured isotopically was three times greater in worms vermicasts than in the underlying soils. Lee (1985) suggests that the passage of organic matter through the gut of the worm results in phosphorus (P) being converted to forms which are more bio-available to plants. This is done partly by the worm's gut enzyme 'phosphatases' and partly by the release of phosphate solubilizing microorganisms in the worm cast.

(xi) *Worms stimulate high levels of beneficial and biologically active soil microbes:* Earthworms host millions of beneficial microbes (including the nitrogen fixers) in their gut and excrete them in soil along with nutrients nitrogen (N) and phosphorus (P) in their excreta i.e. vermicast. The nutrients N and P and the intestinal mucus excreted by worms are further used by the microbes for multiplication and vigorous soil remediation and fertility improvement action. Teotia et al. (1950) reported a bacterial count of 32 million per gram in fresh vermicast compared to 6-9 million per gram in the surrounding soil. The mycorrhizal fungi stimulated and encouraged by the earthworms transfer phosphorus by increasing solubilization of mineral phosphate by the enzyme phosphatase. Morgan and Burrows (1982) showed that the number of beneficial bacteria and 'actinomycetes' contained in the ingested material increased up to 1000-fold while passing through the gut. A population of worms numbering about 15,000 will in turn foster a microbial population in billions in soil.

(xii) *Worms secrete plant growth hormones:* Neilson (1965) reported the presence of 'plant growth substances' in earthworms. Tomati and Galli (1995) also reported that worm worked soil and compost contained the growth promoting hormone 'auxins' and flowering hormone 'gibberlins' secreted by earthworms.

(xiii) *Worms protect plants against various pests and diseases*: There has been considerable evidence in recent years regarding the ability of worms to protect plants against various pests and diseases either by suppressing or repelling them or by inducing biological resistance in plants to fight them or by killing them through pesticidal action. The actinomycetes fungus excreted by the earthworms in their vermicast produces chemicals that kill parasitic fungi such as *Pythium* and *Fusarium*.

5.18. Earthworm Conservation[2]

Threats

i. Rapid pace of habitat destruction/alteration
ii. Spread of peregrine/exotic species and their negative effects on native species
iii. Widespread intensive use of toxic pesticides in certain areas
iv. Lack of interest among conservations
v. Lack of social/political interest to foster conservation of earthworm biodiversity
vi. Serious deficiency in knowledge of abundance and distribution of native species.

Research Needs

i. Training of taxonomists
ii. Survey and inventory; Regional keys
iii. Life history studies of native species
iv. Contribution of native species to litter decomposition, nutrient cycling, etc.
v. Utilization of native epigeic species in vermicomposting
vi. Role of native species in agricultural soils
vii. Competitive interaction between native and exotic species.

[2] Source: Lecture delivered by Dr J.M Julka in Training Workshop on *Inventorying Soil Biota* 17-23 October, 2005 at CSKHPKW. Palampur; YSPUHF, Nauni and Zoological Survey of India, Solan (HP), India.

Chapter 6

ORGANIC FARMING FOR QUALITY MEDICINAL PLANTS' PRODUCTION

The word "organic" means originating from a living thing and farming means production management with the philosophy of organic conversion from the conventional system to make the production system quality assured and free from all contaminants and pollutants. Organic conversion is not just to replace fertilizers and pesticides with organic manure and biopesticides but it is an organic dynamic process for making healthy soil and ultimately a vital living system of the world. Organic farming is the most widely recognized alternative farming system in the agricultural production context. It is a productive system which avoids or largely excludes the use of synthetically compounded fertilizers, pesticides, growth regulators and livestock feed additives. To the maximum extent feasible, the system relies upon crop rotations, crop residues, animal manures, legumes, green manures, off-farm organic wastes including soil humus, mechanical cultivation, mineral-bearing rocks and aspects of biological pest control to maintain soil productivity and tilth, to supply plant nutrients and to control insects, weeds and other pests.

6.1. ORGANIC FARMING AND HERBAL MEDICINE

In the herbal sector, the processing of organic produce has become inevitable due to increased market demand for standardized and quality products. Several studies revealing the presence of pesticide residues, heavy metals and other chemicals in food, medicinal herbs and cosmetic products.

.About one million people suffer from pesticide poisoning and 20,000 die every year due to the toxic effects of chemicals worldwide but the crop loss due to pest is still much higher, to the tune of 3 to 4 billion dollars. In order to avoid the deleterious effects of synthetic chemical fertilizers and heavy metals like lead, cadmium etc., alphatoxins and pesticides, organic cultivation is needed as an alternative to provide an ecologically safe method of farming.

In recent years, there has been an overwhelming response to the organic produce in the world which is growing by 20% per annum. In the medicinal and herbal sectors this is absolutely necessary as all the major European, Japanese and North American buyers are demanding chemical-free raw materials and products. Organic cultivation of medicinal plants is the need of the hour. Therefore, national and international level research efforts are needed to facilitate the successful adoption of organic cultivation and production by farmers and collectors. The approach has to be easy for the farmer, collector, local traders and processors, with simple technologies at an affordable cost.

6.2. Global Market Trend

There is a worldwide market for organically grown, processed and certified herbal products. A coordinated national program and local actions are needed to promote the successful adoption of organic cultivation by poor and marginalized farmers. South Asia, with its rich agricultural diversity stretched across 15 different agro-ecological zones and traditions of agricultural practices, has tremendous potential in practicing organic cultivation and production of medicinal plants, products and services. Most South Asian small and mid-size farmers, especially in the hills and rain-fed areas, practice traditional practices and are organic producers by default, as they cannot afford even the subsidized chemical inputs.

Wide varieties of organic products are sold worldwide and the demand is growing rapidly in Europe, USA, Japan and Australia. India is already exporting a wide range of organic products like tea, spices, cotton, rice etc. Even the Indian domestic market is growing, though it is still very small. In recent years, as several thousands of farmers have switched over to organic farming, the organic production in India has considerably increased. Though one should not expect miracles from organic agriculture, it can be a realistic and promising option, especially for marginal and resource-poor farmers. For gaining consumers' confidence, valid certification is an essential prerequisite. It is also important for marketing, especially in the export market. Organic inspection and certification are complex tasks, and regulations and accreditation criteria are manifold. Only nationally and internationally well recognized accredited certification agencies would facilitate access to both export and domestic organic markets.

Internationally, the demand for organic medicinal, aromatic and dye plants is growing, with more and more people seeking health solutions from alternative and complementary medicines whose worldwide market is estimated at more than 60 billion US$ annually. The inclusion of cosmetics and pharmaceutical health products will perhaps push the value close to 310 billion US$. Greater awareness of the impact of production methods on the quality of a product will lead to a growing number of consumers choosing ecologically and particularly organically grown herbal products.

6.3. Assessment of Demand and Supply

Based on the data available through published sources and a quick assessment survey conducted by the Group on Pharmacy Linkages, it is estimated that the current annual demand for medicinal plants in India alone is about 0.25 million tons and it is growing at a rate of about 15-20% per year. It is also estimated that only about 10 medicinal plants are required by the industry in bulk (1,000 to 10,000 tons), which are mainly obtained from the wild, causing resource depletion.

Currently, the global market for herbal products, which includes medicines, health supplements, herbal beauty and toiletry products, is estimated at around US$ 62 billion. The global market for herbal medicines alone is estimated to be around US$ 5 billion, growing at a rate of 30% to 40% annually and is expected to reach US$ 16 billion by 2005. It is estimated that China accounts for over 40% of this trade, while India's share is estimated to be less than 2%. India, with a strong knowledge base in the traditional systems of medicine and vast natural resources, is in a unique position to move ahead and capture the global market for herbal products. India's comparative advantage for producing herbal medicines for export has not been exploited fully.

India and China are the world's largest producers of medicinal plants and herbal medicines. India, which supplies about 12 percent of the medicinal plants used by the European herbal medicine market, has over 7,500 species of medicinal plants. These serve the health needs of 60 to 80 percent of the country's population, including rural women, tribal communities and local village communities. More than 5,000 small-scale pharmacies operate throughout India, producing traditional herbal medicines with an annual turnover estimated at over US$ 250 million.

6.4. GROWING IMPORTANCE OF CERTIFICATION

Proper inspection and certification systems are essential for developing internationally acceptable organic products. The development and implementation of local or national legislations and policy framework are necessary to support a certification system which is affordable, comprehendible and implementable. The focus should be to put in place a support system to the rural-based production sector, through contracts and joint ventures with national and international certification and accreditation bodies from potential importing countries. Specialized certification and quality assurance systems will be required and implemented with market and industry partners for different products and market sectors. The cost of a typical certification may average 1 to 2% of the final producer price for medium size farms, but certification costs, distribution and certification schemes will have to be adapted to specific local conditions. Although some existing certification schemes are starting to certify specific non-wood forest products, there is considerable variation in what is being certified and some misunderstanding exists as to what certification of a Non Wood Forest Produce (NWFP) actually means. A common assumption is that certification of an NWFP is a guarantee that it has been produced in a sustainable way. There are four different certification approaches:

6.4.1. Certification of origin**:** This is used for a variety of products, including food products. It guarantees that a given product comes from a certain region or area - not necessarily that it has a certain standard or quality.

6.4.2. Product quality standards**:** Organic certification is used for an increasing number of products from food to textiles. It certifies a full production sequence and criteria for organic agriculture. As medicinal and aromatic plants (MAPs) are gathered in forests and their production does not entail the use of fertilizers or pesticides, they are usually considered organic products by definition. However, it does not guarantee that these products have been obtained from sustainable managed forests.

6.4.3. Social certification**:** This involves the documentation of certain social aspects of production, assuring that the labor conditions for production are acceptable and that the benefits are equitably distributed to those involved in production.

6.4.4. Forest management certification**:** Forest certification schemes based on the standard of forest management are now being extended to include MAPs. This is much more complex than simply certifying for timber, as the same forest area may have to be assessed for one or more MAPs, which can have different requirements. It is possible, for example, for a forest to be managed for timber in a sustainable way while, at the same time, its NWFP resources are being over-harvested, and *vice versa.*

6.4.5. Recommendations**:** In order to augment both the quantity and quality of MAPs, the following recommendations are suggested for consideration and implementation by South Asian governments, civil society groups, community-based organizations and private sector concerns.

6.5. Policy Reforms

1. Enabling policy and institutional arrangements.
2. Focus on environment and biodiversity conservation, especially forests, wastelands, botanical gardens, sacred groves etc., as these continue to be the primary habitat for medicinal plants.
3. Government institutions need to be strengthened to develop and regulate MAPs resources and, at the same time, foster their sustainable development and conservation.
4. Link the management and conservation of medicinal plants (and other non-timber forest products) with the commercial development of these resources.
5. Move beyond the demonstration/pilot projects towards comprehensive conservation, use, benefit sharing and sustained human health.
6. Single-window system under an apex body to achieve better coordination and minimize duplications.
7. Developing a computerized database on medicinal plants of the region with as many parameters as possible.

6.6. Organic and Quality Production

1. Development of a simple package of practice for organic cultivation and collection of medicinal plants.
2. Quality control and implementation of standards in production processing systems.
3. Formalizing and organizing market, trade and enterprises through participatory processes.
4. Linking traditional medicine (TSM) with modern medicine (MM) and modern science (NARS).
5. Primary health and agricultural extension agencies should strengthen extension efforts to the MAP collector and farmers.
6. Research institutions need to improve basic knowledge about sustainable harvesting methods, *in situ/ex-situ* cultivation practices, primary processing, marketing and dissemination of information.
7. Conservation agencies and NGOs should promote conservation of vulnerable species and ecosystem at the grass-roots level by creating proper incentives.
8. Community organizations need to develop, adopt and transfer sustainable collection and management practices on common lands.

6.7. Marketing, Certification and Capacity Building

1. Short-term and long-term policy, planning and programming through multi-stakeholder consultations.
2. Profitable community-based and private micro-enterprises should be set up for processing, certification, transporting, and marketing.
3. Proper certification systems need to be developed and practiced to ensure sustainable harvesting, international quality and standards and organic properties.
4. Consumers need to be educated on this issue so that they choose products with the sustainable harvesting and organic/ecological cultivation labels.
5. Capacity building both of the individuals and institutions, especially at the grass-roots level.

CONCLUSION

The policy for medicinal plants in South Asia should be people-centered, livelihood-focused and biodiversity-enriching and based on the long-term vision of providing equitable access, and a fair share of benefits and conservation through sustainable use to the local people. The underlying principles of policy making should be as follows: a) assignment of roles and responsibilities to different stakeholders in a manner commensurate with the demand of their services, institutional capacity and track record of performance in past projects; b) monitoring and evaluation of the performance of agencies should be evidence based and with verifiable progress markers and outcomes; c) the focus should be as much on process as on methodology to achieve the policy objectives; d) actions and activities of various stakeholders in a time bound manner need to be defined and monitored; e) as the policies and their proper implementation are more critical to the poor and local community, the outcome and relevance of any policy reform in MAPs sector should be evaluated based on a tangible improvement to rural livelihoods and an improved quality of biodiversity. There are no simple and easy answers in complex policy debates and discussions, but the ultimate actions of the people whose livelihood is dependent on these resources - the Van Panchayats, Forest User Groups and Joint Forest Management committee functionaries who have undertaken the responsibility for protecting both the forests and livelihoods, and the district and state level government officials in policy and institutional arenas - will help us to understand the nature and extent of the problems and their possible policy solutions.

Chapter 7

CERTIFICATION OF ORGANIC PRODUCTS

Organic farming is both an agricultural philosophy and a farm management system. Contrary to what a lot of advertising would have people believe, organic farming is not just throwing a bunch of seeds on a field and coming back later in the year to harvest what's left. Organic farming is using the very latest in technology, applied to current research, to fulfill the principles of good soil husbandry to which our forefathers adhered.

Organic farmers are aware of the damages done to the soil by applications of harsh chemicals. They believe that in agriculture, to remain productive, the soil must be cared for, not abused. Organic farmers not only wish to avoid the use of chemical fertilizers and pesticides that can cause damage to soil and wildlife, and create toxic side effects in a variety of ways, but they are also very much concerned about the prevention of erosion, and the addition of humus and other organic matter to the soil to improve its fertility, as well as the preservation of small family farms, localized marketing of food, energy conservation and proper nutrition.

Different forms of organic farming in the developing world have complex layers of meaning, perception, institutional involvement and farmers' interests associated with these different forms. It stands with a brief analysis of the most visible and quantified element of organic farming - formally certified organic production. This is a rapidly expanding sector which is proving to be highly successful in increasing returns to farmers through meeting growing consumer demand for organically (and often ethically) produced food.

Certified organic farming in the developing world largely conforms to Northern experiences in that farmers adopt a set of standards and in return normally receive higher prices for their produce. In a world where the terms of trade are often stacked against primary producers this is undoubtedly a benefit. Organic production practices may lead to higher yields, in particular when the starting point has been extensive and low input conditions. However, it generally tends to reduce exposure to toxic chemicals, and risk of crop failure, reduces outgoings and exposure to debt, builds environmental resilience and can build on local knowledge and socio-cultural practices. The widespread existence of non-certified organic farming suggests that these benefits may be sufficient in themselves to attract farmers, even those without access to (or interest in) export production. In this chapter we attempt to assess these benefits of organic farming from three different perspectives: those of classical agronomy (focusing on yields and soil fertility), of agricultural economics (focusing on net returns) and of human ecology, which incorporates the approaches of both multi-functionality and livelihoods.

7.1. Certified Organic Farming in the Developing World

Each year the International Federation of Organic Agriculture Movements (IFOAM) publishes a statistical handbook about organic farming. In recent years these have shown a meteoric rise in the number of certified organic farmers and the amount of land under organic management. Analysis of these figures provides a useful starting point for understanding the status and trends of certified organic farming in the developing world. In overall terms, organic farming is highly developed in Latin America, with almost 6 million ha of certified organic land, constituting 24% of global certified organic land. Three Latin American countries - Argentina, Brazil and Uruguay - are among the top ten countries in terms of total certified land. By contrast, Asia and Africa have relatively less certified organic land, 0.9 and 0.3 million ha, respectively, accounting for 3.7% and 3% of the global total.

In Asia, Sri Lanka is a leading producer, with 0.7% of its cultivated land certified as organic; several other Asian countries (Indonesia, Kazakhstan and Ukraine) have recently entered the organic market with a substantial production base, but because of their size the proportions remain relatively small.

The certified organic production occurs in a context in which a formalized relationship predominates. The global market sets the terms of reference, efficiency, what is (and what is not) organic, provides consumers with guarantees about production methods, producers with premia (often substantial) and prevents free riders. But these figures are a social construction, reflecting participation in global markets that require the setting of standards certifications and inspections and which provide the resources to do this. This reliance on international trade is recognized as a limiting factor and there is a growing move to develop national and local markets. These have gained particular importance in parts of Latin America and serve to link producers with local consumers either directly through farmers markets etc., or indirectly through supermarket chains. These are very important developments in terms of generating better returns for farmers and raising the profile of organic production in domestic markets. Their emergence is an important means of developing organic farming as a commercial activity in many countries as they provide an important 'third way' in which organic farming can be both commercially viable and achieve a distinct local or national profile.

7.2. Explicit Organic Approaches

The IFOAM (The International Federation of Organic Agriculture Movements) membership directory provides one avenue to identify a number of organizations that explicitly identify themselves as organic but whose main concern appears to be with local development priorities. It is interesting to compare these membership figures (Table 1) with the figures for certified organic land. Latin America accounts for 9% of IFOAM's membership (compared to 21% of certified organic land), Asia accounts for 19% of IFOAM's membership (compared to approx. 4% of certified organic land) and Africa 7% of IFOAM's membership (compared to approx. 1% of certified organic land). Some countries have many IFOAM members even though their formal certified organic sector is small. These include Kenya, Senegal and Benin in Africa, India, the Philippines and Nepal in Asia. Equally, there are countries with IFOAM members where there is no record of any certified production. This includes Togo, the Democratic Republic of Congo and Reunion in Africa, Malaysia, Iran, Singapore, Syria, Taiwan and the UAE in Asia and Trinidad and Tobago in Latin America. If we look at these organizations, it is evident that many are development oriented NGOs, with relatively little export orientation. This supports the notion of a world of organic farming removed from the imperatives of the global economy.

One such example is the Kenyan Institute of Organic Farming (KIOF) widely regarded as a pioneer training center. It has trained more than 16,000 farmers in organic techniques over the past 20 years and is visited by farmers and advisors from all over East Africa. A recent letter from KIOF's Director summarized their achievements - *raison d'etre:*

> The emphasis of the training programme is on organic farming for poverty alleviation and household self-sufficiency. Adopting organic farming technique leads to substantial increase in productivity. It also leads to an increase in the diversity of food produced and generates surpluses for sale. Other benefits include savings in expenditure on chemical inputs and health benefits from not being exposed to these. Enhanced food security remains the most visible impact of organic farming among small-scale low income groups. - John Njoroge

Table 1. Members of the International Federation of Organic Agriculture

No. of IFOAM Members	Latin American N=61 (9.5%)	Africa N=45 (7%)	Asia N=125 (19%)
> 30			India (35)
> 20-30			China (25), Japan (21)
> 10-20	Argentina (17)		
> 5-9	Chile (8), Mexico (8), Brazil (6), Peru (5)	Egypt (9), Kenya, Togo (8), South Africa (7), Uganda (6), Benin, Senegal, South Africa (5)	Philippines (9), Turkey (8), Nepal (5)
>5	Costa Rica, Ecuador (4), Bolivia, Guatemala, Nicaragua, Columbia, Cuba, Trinidad, Tobago (1) and Paraguay (2)	Burkina Faso, Dem., Republic of Congo, Ghana, Ethiopia, Cameroon Malawi(3), Mozambique, Zambia (1), Madagascar, Reunion Tunisia, Zimbabwe(1)	Malaysia, Israel(3) Pakistan, Taiwan, Iran, Sri Lanka, Bangladesh, Indonesia, Korea, Palestine, Singapore, Syria, United Arabic Emirates (1)

Source: Global Development of Organic Agriculture: Challenges and Prospects (Eds. N. Halberg, H.F. Alroe, M.T. Knudson and E.S. Kristensen), CAB Publication, New Delhi, 2002.

7.3. Agro-Ecological Approaches

Alongside organizations that formally align themselves to the organic movement, there are a number of other organizations and movements that share very similar approaches to agriculture in their design. These include biodynamic agriculture, permaculture, natural farming, bio-intensive, eco-agriculture and agro-ecology. While farming according to these different styles may not always meet internationally recognized organic standards, the approach and philosophy that they employ is largely compatible with the organic world view and relies on similar agronomic practices. Some of these movements have played a significant role in developing organic approaches in the developing countries. The biodynamic movement has played a key role in developing the organic movement in Egypt and the permaculture movement has launched a host of training initiatives in Zimbabwe.

Low External Input Sustainable Agriculture (LEISA) has been promoted for around 20 years. The movement largely came about as a result of the failures or inappropriateness of the green revolution's approaches to smallholder farming systems. LEISA stresses the importance of developing appropriate technologies and farming systems. Through participatory approaches they promote the optimal use of natural resources and local processes and if necessary, the safe and efficient use of external inputs.

At the same time there are a number of International Research Institutes (IRIs) that have programs or elements of their programs that also appear to be consistent with the organic approach. One striking example is the Biological Control Centre in Cotonon, Benin (part of the International Institute for Tropical Agriculture). This center has developed several programs that rely on organic principles (e.g. post-predator relationships) or the development of locally produced myco-pesticides (www.lubilosa.org/INDEX.htm). However, there is no explicit recognition of the 'organic' nature of their work and collaboration between the organic movements.

Finally, we come to the most problematic area, that of resource-poor farmers' farming without the use of agrochemicals. This approach is often driven by poverty and lack of resources and characterized by low outputs and unsustainable practices, rather than by the conscious adoption of organic farming techniques or acceptable levels of productivity (FAO, 2003). Studying this group focuses attention on what constitutes organic farming and the extent to which traditional practices conform to organic designs. One recent report that examined the extent and potential of organic farming in sub-Saharan Africa identified that: *Isolated organic techniques are sometimes practiced but there is a general lack of an integrated approach to soil fertility and crop protection management and under exploitation of the full range of techniques that would maximize the benefits of locally available natural resources* (Harris et al., 1998).

It is exceptionally difficult to make a general assessment of the extent of organic farming practices in traditional farming systems. There are some shining examples of locally developed farming systems that make extensive use of available natural resources, maintain soil fertility and generate a wide range of products to meet both domestic needs and sell surpluses on local markets.

The agro-ecological approaches raise three critical points. First, they show that far more farmers use organic practices than the figures for certified organic farming suggest, although it is difficult to ascertain how many because of blurred boundaries with agro-ecological approaches and traditional farming. Second, they show that a number of organizations are involved in doing research in or disseminating organic practices, although many do so without explicitly presenting their activities as organic. Third, there may be a potential for strengthening organic approaches if the appropriate dissemination strategies can be put in place.

In the context of food security it is further important to remain aware of the key role that agricultural systems themselves play in determining entitlement to food. Systems that depend upon the sustainable use of locally available natural resources and farmers' knowledge and labor are far more likely to meet the needs and aspirations of resource-poor farmers than those which require costly or scarce external inputs. In this respect organic farming is a technology that, to quote Gandhi, 'puts the last man first' and as we have come to realize since Gandhi's day, the last man is more often than not a woman.

7.4. Analysis for Organic Farming from a Small Farmer's Perspective

Strengths:

a. Often close to existing practice.
b. Fit with risk-averse strategies of farmers.
c. Can valorize local resources, creating new uses for them.
d. Reduces risk of exposure to harmful pesticides (which are often improperly used).
e. Reduces exposure to the debt required to purchase inputs.

Opportunities:

a. Possibility for greater security and selling surpluses locally.

b. Can create new on-farm income generating opportunities.
c. Possibility for premier prices if certified.
d. Opportunities for greater social contacts through project meetings etc.

Weaknesses:

a. Organic farming can involve substantial additional labor input, especially in the initial phase.
b. Investment of time and other resources of farmers might not seem worthwhile if benefits are not apparent, if the farmer has more pressing problems or lack secure tenure.
c. Lack of knowledge and readily available information.
d. Poor (or nonexistent) extension services, which often promote conventional approaches.
e. Lack of local markets.
f. Cost and implications involved in certification.
g. Low literacy rates hamper record keeping.

Threat:

a. Belief in modernization peer group ridicule.

7.5. Certification of Organic Produce

The food market, world over, has been virtually flooded with products complying with varying standards. Presently, the food products are required to comply with standards prescribed under different regulations related with quality parameters applicable in each country. Despite most of the regulations already in place for controlling the quality, still many products are sub-standard. The consumer as a result is unsure and doubtful about the authenticity of even existing products. Certification in the case of organic agriculture, thus, means having the farms and farming methods inspected by an agency that would ensure the compliance with a prescribed standard. This certification should ensure a hundred per cent satisfaction of the consumers of organic food.

A wide variety of organic products are now being sold worldwide and the supply is keeping pace with the demand that is growing rapidly. For gaining consumers' confidence, valid certification is an essential prerequisite for organic marketing, especially in the export market. Organic certification is also an indication of positive management systems.

Organic inspection and certification are complex tasks and regulations and accreditation criteria are manifold. A well designed and recognized system can only ensure certification. Another important aspect of certification is its affordability to all farmers, including small and marginal ones.

For a credible certification system, the certification agency must be accredited. It should also be independent and must follow a set certification procedure. The cost of certification should also be reasonable enough that it does not force the consumer to pay an extraordinarily high premium for this quality certification.

Organic consumers look for certain attributes attached to organic products; i.e. safer (no pesticides or other harmful chemicals), more nutrients, higher quality and environment friendly. The world over, consumers are paying significant premiums for certified food because they trust the integrity of recognized organic certification marks.

The biggest difference between conventional and organic food production is the requirement for the certification. Organic certifiers represent the consumer and provide the guarantee that the food is grown and processed according to acceptable standards.

Different certification agencies offer their certification services for different organic production standards. Both product and process certification services are offered by them. Some agencies offer both individual as well as group certification schemes. Generally, producers of organic food are categorized as:

a) Food producers - cereals, tea, spices, etc.
b) Food processors - milling units, fruit and vegetable processing units
c) Livestock farmers
d) Aquaculture
e) Herb collectors
f) Honey producers
g) Inputs, e.g. compost etc.

In addition, some agencies also offer certification services for textile production.

7.6. Certification Process

The certification process adopted by each certification agency is predetermined and must comply with the provisions laid down by accreditation agencies. While process parameters for certification are decided, they are generally designed to avoid any future conflict and to maintain a high degree of competence and to ensure uniformity and consistency. The certification process is also designed for primary production and non-primary production.

Primary production certification for primary producers can take some time as prescribed under the certification norms. Some certification agencies also prescribe minimum duration; for example, SGS takes a minimum of 12 months to as long as 36 months to issue an organic certificate.

The first step for certification begins with an application for certification. Each agency normally has a prescribed format giving details of terms and conditions along with the standard application form. In many cases, a proposal for certification giving complete details of different stages of certification and the costs involved is also sent to the applicant.

On receipt of this application, information required for preliminary screening is sought from the applicant. This can include a map, details of operations, management plans, soil test reports and an affidavit declaring the last prohibited input use. Once the required information has been collected by the certification agency, it may officially register the applicant for certification. Immediately on registration, the first audit (also known as registration audit) is carried out by the certification agency.

In this audit, if everything goes well with the first two documents, the registration is confirmed and the time period known as the "conversion period" starts from the day of the visit. During the inspection visit, a list of activities, generally carried out by the inspectors, is as follows:

a) Discussion with the operator about the practices, standards, and methods followed for production of the item under certification.
b) Discussions with the laborers and others involved in the production process.
c) Review of records maintained by the operator and corrective measures adopted based on the previous suggestion.
d) Visit to farm/production site and make observations about the organic and non-organic farms, and other premises like office, godowns, processing units etc.
e) Discussions with other farmers of nearby area or some responsible persons to cross-check/verify the above information.
f) Collection of samples/representative samples of materials used during the production process randomly from many places and mixed

thoroughly. In the case of perishable items, the emphasis shall be laid on checking the inputs rather than final products.

g) Inquiry about problems faced by the operator and possible suggestions.
h) Three sets of samples are made, sealed and signed by the monitor/personnel visiting the farm, operator and other responsible person. One is retained by the operator and the other two are sent to the regional coordinator/ head office as decided from time to time for further testing.
i) A report is prepared by the monitors/personnel visiting the farm after completion of the inspection, signed, as in the case of the samples, and one copy is retained by the operator. This also includes the problems faced by the operator and the suggestions given. Such problems are highlighted for which solutions are expected.
j) At the end, the operator is informed of the tentative date for next visit.

7.7. Special Provisions for Monitoring -Surveillance

Audits by inspectors, also known as unannounced inspection visits, are also carried out. During such visits either a few activities carried out recently or those being carried out during the visit are verified according to the conversion plan and the norms are mutually agreed between the operator and the certification agency. The records of each visit are maintained meticulously and are also supervised by senior functionaries of the certification agencies.

Testing of samples can also be carried out to detect the use of any prohibited substance or materials. In some cases, on the basis of compliance with organic standards achieved by the operator, the certification agencies can issue "full conversion certificates" and in some conditions, certificates are issued stating "in conversion to organic".

Non-primary production certification covers processors, exporters, input product manufacturers and suppliers. The certification-approval process for such operators can be completed in shorter durations of 3-6 months. The operators are required to submit complete details including operating procedure, protocols, inputs etc. Once the documentation is complete, the audit is carried out by an inspector designated by the certification agency for compliance with standards.

7.8. ORGANIC GROUP CERTIFICATION

Recently introduced, this procedure allows groups of farmers - or marketing agencies such as traders dealing on behalf of many farmers - to sell their products as Certified Organic, according to the IFOAM criteria for certification "Small Holder Group has been defined as an organized group of small-scale producers with similar farming and production systems."

Generally, in group certification, internal monitoring and control mechanisms are introduced for maintaining compliance with standards. Random sampling of some members is carried out by the inspectors from certification agencies during audits. Group certification saves on costs for individual operators. To qualify for group certification certain agencies insist on qualifying criteria such as:

a) Group comprised of producers with similar farming systems and production from the same area.
b) Have some form of coordinated marketing activity to enable oversight of the product's flow.
c) Group is large enough to support a viable internal control system that assures compliance of individual operators in an objective manner.
d) Groups may include producers with small processing and storage units.

In such cases, the criteria for compliance are divided into two categories i.e. at group level and at individual operator level. The person responsible for the group (also known as group manager) is made responsible for:

a) Communications with the auditors and with primary producers
b) Implementing the administrative requirements of the group certification program
c) Implementing management and/or monitoring responsibilities at the producers' level.

In many such cases, the licensee is the group and not the individual operators. Group certification in recent months has started gaining greater acceptance. The very high cost of certification on an individual basis has also contributed to the evolving group certification norms. The recent norms and standards launched by IFOAM have now also included group certification procedures. In some countries small producers up to certain turnover limits are allowed to adopt less rigorous procedures and are exempted from cumbersome record maintenance requirements.

7.9. Acceptance of Prior Certification

In view of every country prescribing its own certification system, standard procedures are being developed for the acceptance of products based on recognition of a certification body other than the country's own certification. As per IFOAM criteria for equivalency, acceptance of the prior certification of operators seeking certification status should be done.

In this regard an initiative by all IFOAM accredited certification bodies is the right step in this direction. One of the IFOAM publications *Ecology and Farming, No. 23, January-April 2000* states: *For the purpose each certification body is expected to maintain a formal register of recognized certification bodies having equivalency. This can be done on recent and adequate evaluation visits either by two bodies granting acceptance or by an appropriate third party. In the absence of an equivalency agreement or contract of recognition, the certification body can also accept previous certification based on document review. For this purpose assessment of the information contained in the previous inspection report last inspection decision and other relevant documents provided by other body can be carried out.*

IFOAM even permits the transfer of certification from one certification body to another, provided both the certification bodies are accredited. A contract has to be prepared for this purpose and the operator can even continue with dual certification. Similarly, joint partnership between two or more certification bodies can also be made but, in any case, the certification decision is not allowed to be sub-contracted.

7.10. Certification Agencies

All over the world, many agencies are accredited for organic certification. Each agency has its own areas of operations and is accredited under different certification standards. Many international agencies are also operating in India. The scope of these guidelines does not permit detailed information about certification agencies operating in India.

However, a brief description of some agencies about which information is readily available is given below as representative examples:

A) *ECOCERT* is an inspection and certification body established in France in 1991 by agronomists aware of the need to develop environmentally friendly agriculture and of the importance of offering some form of recognition to those committed to this method of production. From its creation, *ECOCERT* has specialized in the certification of organic agricultural products. Certification is issued according to ED regulations in the United States, Japan and other countries. ECOCERT SA is accredited according to ISO Guide 65 (European Norm EN 45011) for its international business by COFRAC, the public French accreditation agency. COFRAC is a member of EA (European Accreditation) and IAF (International Accreditation Forum) and is thus an internationally recognized accreditation body. ECOCERT in India has an office at Aurangabad, Maharashtra. A document set containing application form for inspection and certification services for organic products according to EC regulation 2092/91, as amended, has been analyzed and was found to be very user-friendly.

B) INDOCERT qualified as the first Indian agency engaged in organic certification. INDOCERT has strong technical collaborations with two well reputed Swiss Organic Agriculture Institutions: FIBL and bio.inspecta. These agencies assist INDOCERT in efficient management or certification and co-certification for export markets. INDOCERT is also accredited with the National Standards for Organic Production notified under the National Programme on Organic Production by the Spices Board. INDOCERT in India has an office at District Ernakulam, Kerala. It is also offering membership to individuals and organizations engaged in organic agriculture. The application for single producers or group certification has also been simplified by INDOCERT. In India INDOCERT is offering inspection and certification for European regulations that is certification for the export market according to EC - 2092/91 (EU Regulation).

C) IMO Control inspects and certifies the quality control system adopted for organic agriculture in various countries, including India. IMO is also accredited by the Forest Stewardship Council (FSC) for the certification of ecological forest management, and wood products. It is accredited with NSOP-India, Regulation (EEC) No. 2092/91, NOP (USA) etc. and is also authorized to verify the implementation of Private Standards such as NA TURLAND, BIOSUISSE and DEMETER. IMO Control in India has an office at Bangalore. IMO offers its partnership in the field of inspection and certification of organic agriculture products. Many European importers of organic produce have their organic farming operations in different parts of the country certified by IMO.

D) *Naturland e.V.* Association for organic agriculture is a private certification body and an organic farmers' association. Naturland was founded in 1982 in Grafelfing near Munich (Germany). Naturand is registered as a non-profit organization. In Germany 1633 Naturland farmers are cultivating approx 123000 hectares. In 1989 Naturland started its international commitment. Since 1982 Naturland has been promoting organic agriculture throughout the world; it is currently active in 50 countries with its 53,000 farmers and over 500 manufacturers as Naturland partners. It is one of the major organic farming associations. Naturland regards organic competence and social responsibility as interdependent. Social standards are an integral part of each Naturland certified product. Naturland certification standards range from organic food production, organic aquaculture and sustainable capture fishery to processing and trade as well as non-food areas like organic forest management, organic cosmetics and organic textiles. Worldwide, 27000 farmers, mostly organized in smallholder cooperatives, cultivate more than 79300 hectares following Naturland standards. It was also the first German Organic certifier to attain IFOAM accreditation in 1997 and to be accredited according to European Norm DIN EN 450 II/ISO 65 in 1998. Naturland has recently offered a very innovative service fee package for organic certification. Naturland e.V. in India has its office at Gurgaon, Haryana.

E) *Skal International* is a leading international certification agency. The main certification programs offered by the organization include BRC, EUREPGAP, Forestry and Fisheries, Utz Kapeh Koffee and textile certification..Skal International in India has an office at Bandra, Mumbai.

F) SGS-India Pvt. Ltd, SGS-International is one of the oldest inspection, testing and verification companies and operates in more than 140 countries. SGS Organic Production Standard complies with a number of international requirements including EEC regulation; IFOAM; Codex Alimentarius; UK Register; Accredited by RvA -Dutch and APEDA. SGS has a well laid-out certification process for primary production as well as non-primary production. It begins with an organic certification questionnaire (though slightly complicated for a lay person) and follows the sequence of Proposal for Certification, Application for Registration, Registration Audit, Surveillance Audit, In Conversion Audit, Full Organic Audit to Certification to Organic. In India, SGS offers group certification services also. This allows traders who are dealing with many farmers to sell their product as certified organic. In group certification, it allows for a degree of random sampling with consequential cost saving. To qualify for group certification it also insists on a viable internal control system. SGS in India has an office at Gurgaon, Haryana.

G) *Demeter:* Demeter International is the largest certification organization for biodynamic agriculture, and is one of three predominant organic certifiers. Its name is a reference to Demeter, the *Greek goddess* of grain and fertility. Demeter Biodynamic Certification is used in over 50 countries to verify that biodynamic products meet international standards in production and processing. The Demeter certification program was established in 1928, and as such was the first ecological label for organically produced foods. Demeter's "biodynamic" certification requires biodiversity and ecosystem preservation, soil husbandry, livestock integration, prohibition of genetically engineered organisms and viewing the farm as a living "holistic organism". The certification verifies the fulfillment of the standards on behalf of the farmers, which in turn guarantees high quality food products to the consumers. This is rewarded by receiving a higher price for food certified with the "Demeter" label, ranging from 10-30% on average. The origin of Demeter is a cooperative for the processing of products of the biodynamic agriculture created in Berlin, Germany, in 1927. The trademark Demeter was registered in 1928. Demeter was administered by the German agronomist Erhard Bartsch, who also directed the Experimental Circle of anthroposophical (biodynamic) farmers, and who chose the name Demeter jointly with the German chemist Franz Dreidax. Dreidax was responsible for the development of the Demeter criteria and the quality control. Demeter ceased temporarily to exist in 1941 when the Nazi Government dissolved the Union for biodynamic agriculture. It was reestablished in Germany after the Second World War. In 1997 nineteen independent Demeter organizations came together to establish Demeter.

H) *Bio. inspecta:* Bio.inspecta, founded at the end of 1998, has taken over the inspection and certification functions formerly performed by the Control Service of the Research Institute for Organic Agriculture and BIO SUISSE. In 1999, it was accredited by the Swiss Federal Office of Metrology as an inspection and certification body for unprocessed and processed products from organic agriculture.

I) *M. P. Government, Department of Agriculture:* The Department of Agriculture has evolved the complete procedure for the certification of organic agriculture in M.P. It comprises of formats for application by a farmer, recommended by village committee, registration with Senior Agriculture Development Officer and so on. However, this procedure may not be acceptable according to normal organic standards. It is an excellent beginning for the farmers in Madhya Pradesh.

J) *APOF:* A recent entrant in formulating its own Standards for Organic Production based in Bangalore. APOF has also proposed to certify according to "India Organic" standards.

K) *KARM:* A Delhi based agency working with many farmers' groups has also entered into the certification arena. KARM has proposed to become an accredited certification agency according to "India Organic" standards.

Considering the number of farmers likely to take up organic agriculture in India, it is desired that in addition to international certification agencies many more Indian agencies in different parts of the country will really help to promote organic agriculture.

Both the Ministry of Commerce as well as the Ministry of Agriculture should work to develop this basic infrastructure for organic agriculture development in India. Many private initiatives already taken up by NGOs should also be documented. The local infrastructure, by involving local experts, can meet the ever-growing demand for certification agencies in India.

7.11. Certification Procedures

Many farmers are today exploring the possibilities of conversion to organic production management. Organic agriculture in simplified terms is generally defined as chemical-free farming, but in reality it is much more than just this. The organic certification is primarily based on process documentation requiring complete records to be maintained for each operation carried out on the farm. A simplified step-by-step procedure for organic certification is given below:

- The cost of certification quoted by the certification agency is also dependent on the primary information submitted by the farmer. It is, therefore, necessary that each farmer wanting to take up certified organic production must first of all begin with the preparation of basic documents giving complete details of his farm unit.
- One more complexity generally encountered by farmers before their conversion to organic is related to the selection of a certification agency and its consequent accreditation with different Organic Standards. It may so happen that if this particular aspect is not carefully addressed before the selection of a certification agency compliant with particular Organic Standards, a producer may obtain a certification that is not acceptable to a buyer. It is therefore preferable that, right from the beginning, the producer should try to maintain the minimum benchmarks acceptable to a wide range of Organic Standards, even before he approaches a certification agency. Ideally, once the national standards have been evolved and have been accepted on a reciprocal basis, the producer would comply with only his own country's standards.
- The premium price of commodities associated with organic agriculture has been attracting many farmers. However, before a final decision is taken, it would be better for a producer to carry out a detailed analysis of his present situation compared to the new situation under organic management. Organic agricultural practices require a complete transformation from conventional practices being carried out by the producers. It also requires the on-site management of inputs, adoption of agronomical practices, processing and even direct marketing of organic produce by the farmers according to the

standard norms. This kind of change requires careful and detailed analysis before committing to adopt organic agriculture.

- The record maintenance requirement for organic certification is generally prescribed by the certification agency. The formats according to the standard norms are also provided at the time of registration. It has been observed that many individual farmers have found it difficult to maintain such elaborate records. One solution to this problem, and especially in the context of prevalent illiteracy, is the group basis of organic certification.
- Once a farmer/producer or a group of farmers decides to take up certified organic production, they are required to inform the selected certification agency. In response, the certification agency will first send a questionnaire to the applicant. Each agency has its own questionnaire.
- Based on the information received by the certification agency, it will prepare a proposal for certification. This proposal will tell the party the different stages in which it is going to work with them and how much it will cost. Many agencies at this stage also complete a few other documents such as a complete basic record of the farm unit and also an agreement-cum-declaration by the party. At this stage many certification agencies also issue a registration certificate that indicates the status of a farm unit (being registered with them) as a registered operator.
- The next step is a registration audit or inspection audit. It is the first audit after signing the contract between a producer and a certification agency. Each agency has its own procedure for conducting this audit. Generally, during this first inspection audit a micro-planning exercise is also carried out. In this exercise every activity that would be carried out by the operator for the production of organic produce is discussed and mutually agreed between them. This also includes the documents - records that would be maintained by the operator.
- From this point onward, surveillance audits are carried out by the certification agencies. In this audit both pre-arranged as well as surprise visits can be made to check the compliance with the organic standards.
- Many agencies on completion of a fixed duration may be six months or one year conducting an in-conversion audit. In this audit, if everything has been as stated in the micro-plan and operations are

verified to comply with organic standards, an in-conversion certificate is issued by the certification agency.

- A full organic audit, also known as a final audit, is generally carried out by the certification agency on completion of the conversion period. According to different organic standards, this duration can be a maximum of three years or any duration less than this, provided the certification agency is satisfied with the compliance with the organic standards and its verification.
- At the end, a formal certificate is issued by the certification agency.
- The certified farm operations are then subjected to annual procedures for the continuity of certification.
- In the case of non-primary production activities, generally the procedure for organic certification is quite different and can be completed within 2-3 months also.

7.12. Marketing of Organic Farming Produce in India

The major organic produce in India includes plantation crops like tea, coffee and cardamom; spices like ginger, turmeric, chillies and cumin; cereals like wheat, rice, jowar and bajra; pulses like pigeonpea, chickpea, green gram, red gram and black gram; oilseeds groundnut, castor, mustard sesame; fruits like banana, sapota, custard apple and papaya; vegetable like tomato, brinjal and other leafy vegetables; other crops/products: honey, cotton and sugarcane especially for jaggery (GOI, 2001). The organic products available in the domestic market are rice, wheat, tea, coffee, pulses, and vegetables as well as cashewnuts, cotton, oil seeds, various fruits and medicinal herbs available for the export market.

Wholesale traders, supermarkets and produce shops are the major channels in the domestic market in metropolitan cities and account for only 7.5 per cent of the total organic products market channel for the export of organic products with the exception of tea which is produced and exported by the tea estate! Quality production with traditional methods, low use of chemical inputs in mountain and tribal areas, easy availability of cheap labor, NGO interventions and various types of support provided by the governments are the features of the Indian organic farming scenario.

BIBILIOGRAPHY

Anonymous. *Biotech News Letter*: Vermicomposting empowering SC/ST Aligarh UP. New Delhi, India: Department of Biotechnology, Ministry of Science & Technology Government of India; 2003; 35-38.

Aiyer, K.S.P. (1929). An account of the Oligochaeta of Travancore. *Records of Indian Museum 31,* 13-76.

Al-Dahmani, J.H., Abbasi, P.A., Miller, S.A. & Hoitink, H.A.J. (2003). Suppression of bacterial spot of tomato with foliar sprays of compost extracts under greenhouse and field conditions. *Plant Disease 87*, 913-919.

Arancon, N . Q ., Edwards, C.A., Bierman, P., Metzger, J.D., Lee, S. & Welch, C. (2003). Effects of vermicomposts on growth and marketable fruits o f f i e l d -grown tomatoes, peppers and strawberries. *Pedobiologia 47*, 731-735.

Arancon, N.Q., Edwards, C.A. & Bierman, P. (2006). Influences of vermicomposts on field strawberries-2: Effects on soil microbiological and chemical properties. *Bioresource Technology 97*, 831-840.

Arancon, N.Q., Edwards, C.A., Bierman, P., Welch, C. & Metzger, J.D. (2004). Influences of vermicomposts on field strawberries-1: Effects on growth and yields. *Bioresource Technology 93*, 145-153.

Arancon, Norman (2004). An Interview with Dr. Norman Arancon; In Casting. http://www.vermico.z,9 (2).

Atiyeh, R.M., Dominguez, J., Sobler, S. & Edwards, C.A. (2000). Changes in biochemical properties of cow manure during processing by earthworms (*Eisenia andrei*) and the effects on seedling growth. *Pedobiologia 44*, 709-724.

Barley, K.P. & Jennings A.C. (1959). Earthworms and Soil Fertility: The Influence of Earthworms on the Availability of Nitrogen. *Australian Journal of Agricultural Research 10*, 364-370.

Bather, E. A. (1920). *Pontoscolex latus,* a new worm from Lower Ludhlow beds. *Annals and Magazine of Natural History 9*, 5-10.

Beddard, F.E. (1902). On two new earthworms of the family Megascolecidae. *Annals and Magazine of Natural History 9*, 456-463.

Bengtson, S. & Zhao, Y. (1997). Fossilized metazoan embryos from the earliest Cambrian. *Science 277*, 1645-1658.

Benham, W.B. (1890). An attempt to classify earthworms. *J. Microse Science 31,* 201-315.

Bhadauria, Tunira & Saxena, Krishan Gopal (2010). Role of Earthworms in Soil Fertility Maintenance through the Production of Biogenic Structures. *Applied and Environmental Soil Science* doi:10.1155/2010/816073.

Blakemore, R.J. & Temple–Smith, M.G. Impacts of earthworm on plant production. In: Temple-Smith, M.G. A Pinkard, T. (Eds.). *The Role of Earthworms in Agriculture and Land Management*. Tasmania: DPTF, Launceston; 1996; 83-89.

Blakemore, R.J. Earthworms of south-east Queensland and their agronomic potential in brigalow soils. Australia: University of Queensland, Australia; 1994.

Blakemore, R.J. (1997a). Two earthworms (Megascolecidae: Oligochaeta from Rennell, Solomon Islands. *Journal of Natural History 31*, 17-25.

Blakemore, R.J. (1997b). Agronomic potential of earthworms in brigalow soils of south-east Queensland. *Soil Biology and Biochemistry 29 (3/4)*, 603-608.

Blakemore, R.J. (1999). The diversity of exotic earthworms in Australia – a status report. *Transactions of the Royal Zoological Society,* 182-187.

Blakemore, R.J. (2000). New species of the earthworm genus *Anisochaeta* from New South Wales. *Records of the Australian Museum 52(1)*, 1-40.

Blakemore, R.J. (2001). Finding Fletcher's giant worms – From Burrawang to Budderoo. *Eucryphia, 54,* 6-7.

Blakemore, R.J. (2003). Japanese Earthworms (Annelida: Oligochaeta): A Review and Checklist of species. *Organisms, Diversity and Evolution 3(3)*, 241-244.

Blakemore, R.J. A provisional list of valid names of Lumbricoidea (Oligochaeta) after Easton, 1983. In: Moreno, A.G. & Borges, S. (Eds.). *Advances en taxonomia de lombrices de tierra / Advances in earthworm taxonomy (Annelida: Oligochaeta).* Spain: Editorial complutense, Universidad complutense, Madrid, Spain; 2004; 75-120.

Blakemore, R.J. VermEcology: Cosmopolitan Earthworms. 2nd Edition. Japan: Yokohama, Japan; 2006.

Bohlen, P.J., Scheu, S., Hale, C.M., McLean, M.A. & Migge, S. (2004). Nonnative invasive earthworms as agents of change in northern temperate forests. *Front. Ecol. Environ.* 2, 427-435.

Bossuyt, H., Six, J. & Hendrix, P.F. (2005). Protection of soil carbon by micro aggregates with in earthworm casts. *Soil Biology and Biochemistry 37*,251-258.

Bouche, M.B. Relation d'entre les structures spatiales et Fonctionelles des ecosystems illustrees par le role pedobiologique des verse le terre. In: C. Delamere Beboutteville (Ed.). *La vie Dans les sols. Paris: Gauthiers villars*, Paris; 1971; 189-209.

Bouche, M.B. *Lombriciens* de France Ecologie et Systematique, Paris: Institut National de la Recherche Agronomique, Paris; 1972.

Bouche, M.B. The establishment of earthworm communities. In: J. Satchell (Ed.). *Earthworm Ecology: from Darwin to Vermiculture.* London: Chapman and Hall, London; 1983; 431-448.

Bouche, M.B. (1984). Ecotoxicologie des lombriciens. Ecotoxicologie controlee. *Acta Oecologica 5*,271-287.

Brown, G.G. (1995). How do earthworms affect microfloral and faunal community diversity? *Journal of Plant and Soil 170*, 209-231.

Buckerfield, J.C., Flavel, T.C., Lee, K.E. & Webster, K.A. (1999). Vermicompost in Solid and Liquid forms as a Plant-Growth Promoter. *Pedobiologia 43*,753-759

Canellas, L.P., Olivares, F.L., Okorokova A.L. & Facanha, R.A. (2000). Humic acids isolated from earthworm compost enhance root elongation, Lateral Root Emergence and Plasma Membrane H^+ -ATPase Activity in Maize Roots. *In J. of Plant Physiology 130*, 1951-1957.

Chakraborty, N.K., Pal, T.K., Dey, S.K. & Varghese, Y.A. (2002). Flora of rubber and other plantations in Tripura. Indian *Journal of Natural Rubber Research 15(1)*,103–106.

Chaoui, H.I., Zibilske, L.M. & Ohno, T. (2003). Effects of earthworm casts and compost on soil microbial activity and plant nutrient availability. *Soil Biology and Biochemistry 35 (2)*, 295-302.

Chaudhuri, P.S., Nath, S. & Paliwal, R. (2008). Earthworm population of rubber plantations (*Hevea brasiliensis*) in Tripura, India. *Tropical Ecology* 49(2), 225-234.

Clausen, M.W. (1993). Regnorme. Danmark's Fauna. *Dansk Naturhistorisk Foreing 84*, 1-176.

Coupland, T. Grey. & McDonald, I. Justin (2008). Extraordinarily high earthworm abundance in deposits of marine macro-detritus along two semi-arid beaches. *Mar. Ecol. Prog. Ser 361*,181–189.

Csuzdi, C.S. & Szlavecz, K. (2003). *Lumbricus friend* Cognetti, 1904, a new exotic earthworm in North America. *Northeast Naruralist 10(1)*, 77-82.

Csuzdi, C.S., Zicsi, A. & Misirlioglu, M. (2006). An annotated checklist of the earthworm fauna of Turkey (Oligochaeta: Lumbricidae). *Zootaxa* 1175: 1-29.

Csuzdi, C.S., Pavlicek, T. & Misirlioglu, M. (2007). Earthworms of the Hatay Province, Turkey with description of new species (Oligochaeta: Lumbricidae, Criodrilidaae, Acathodrilidae). *Acta. Zool. A. S. Hungaricae 53(4)*, 347-361.

Darwin, C. *On the origin of species by means of natural selection, or the preservation of favored races in the struggle for life*. London: John Murray, London, UK;1859.

Darwin, C. *The Formation of Vegetable Moulds Through the Action of Worms.* London: John Murray, London, UK; 1881.

Darwin, C. (1838). On the Formation of Mould. *Proceeding Geological Society London 2*, 574-576.

Easton, E.G. (1981). Japanese earthworms: a synopsis of the Megadrile species (Oligochaeta). *Bulletin of the British Museum (Natural History) Zoology 40(2)*, 33-65.

Edwards, C.A. & Arancon, N. (2004). Vermicompost suppress plant pests and disease attacks. In Rednova News: http://www.rednova.com/display/ ?id =55938

Edwards, C.A. & Burrows, I. The Potential of Earthworms Composts as Plant Growth Media. In: Edward, C.A. & Neuhauser, E.F. (Eds.). *Earthworms in waste and environmental management.* The Netherlands: SPB Academic Publishing, The Hague; 1988; 21-32.

Edwards, C.A. & Fletcher, K.E. (1988). Interaction between earthworms and microorganisms in organic matter breakdown. *Agriculture Ecosystems and Environment 24*, 235-247.

Edwards, C.A. & Lofty, J.R. *Biology of earthworm*, 2nd Ed. London: Chapman and Hall, London; 1977.

Edwards, C.A. & Bohlen, P.J. *Biology and ecology of earthworms*, 3rd Ed. London: Chapman and Hall, London; 1996.

Edwards, C.A., Domínguez J. & Arancon, N.Q. The influence of vermicomposts on plant growth and pest incidence. In: Shakir, S.H. & Mikhail, W.Z.A. (Eds.). *Soil Zoology for Sustainable Development in the 21st Century*. Egypt: Self-Publisher, Cairo, Egypt: 2004; 397-420.

El-Duweini, A.K. & Ghabbour. S.I. (1965). Population density and biomass of earthworms in different types of egyption soils. *Journal of Applied Ecology* 2,271-287.

Fragoso, C., Brown, G. & Feijoo, A. (2003). The influence of Gilberto Righi on tropical earthworm taxonomy: the value of a full-time taxonomist. *Pedobiologia 47*, 400-404.

Frankenberger, W.T. & Arshad, M*Phytohormones in Soils: Microbial Production and Function*. New York: Marcel and Deckker Pub., New York; 1995.

Frelich, L.E., Hale, C.M., Scheu, S., Holdsworth, A.R. & Heneghan, L. (2006). Earthworm invasion into previously earthworm–free temperate and boreal forests. *Biological Invasions 8*, 1235-1245

Gates, G.E. (1937). Indian earthworm: the genus *Pheretima*. *Records of Indian Museum 39*, 175-212.

Gates, G.E. (1938a). Indian earthworms: the genus *Eutyphoeus*. *Records of Indian Museum 40*, 39-119.

Gates, G.E. (1938b). Indian earthworms: the genus *Lampito marutiii* Kinberg with a note on Indian species of woodwardiella. *Records of Indian Museum 40*, 423-429.

Gates, G.E. (1939a). Indian earthworms: *Nelloscolex* sp. *Records of Indian Museum 41*, 37-44.

Gates, G.E. (1939b). Indian earthworms: contributions to a revision of the genus *Eudichogaster*. *Records of Indian Museum 41*,151-218.

Gates, G.E. (1945). The earthworms of Burma. *Asiat. Soc. Beng. 11*, 54-91.

Gates, G.E. (1947). Earthworms of the Allahabad sector of the Gangetic plain. *Proc. Natn. Acad. Sci India 17*,117-128.

Gates, G.E. (1958). On some species of the oriental earthworm genus *Pheretima* Kinberg, 1867, with two species reported from the Americas. *Amer. Mus. Novitates*, 1-30.

Gates, G.E. (1959). On a taxonomic puzzle and the classification of the earthworms. *Bull. Mus. Comp. Zool. Harv. 121*, 229-261.

Gates, G.E. (1972). Burmese earthworms: An introduction to the systematics and biology of Megadrile Oligochaetes with special reference to Southeast Asia. *Trans. Am. Phil. Soc. 62 (7)*, 1-326.

Gates, G.E. (1982). Farewell to North American Megadriles. *Megadrilogica 4*,12-77.

Glaessner, M.F., Priess, W.V. & Walter, H.R. (1969). Precambrian columnar stromolites in Australia: Morphological and stratigraphic analysis. *Science 164*, 1056-1058.

Gobat, J.M., Aragno, M. & Matthey, W. *The living soil: Fundamentals of soil science and soil biology*. USA: Science Publishers, Enfield (NH), USA; 2004.

Graff, O. Preliminary experiment of vermicomposting of different waste materials using *Eudrilus eugeniae* Kinberg. In: Appelhof, M. (Ed.). Proceedings of the workshop on *Role of Earthworms in the Stabilization of Organic Residues*. USA: Malanazoo Pub. Michigan, USA; 1981; 179-191.

Graff, O. & Makeschin, F. (1980). Beeinflussung des Ertrags von Weidelgras (Lolium multiflorum) durch Ausscheidungen von Regensurmeh dreier verschiedener Arten. *Pedobiologia 20*,176-180.

Grappelli, A., Tomati, V., Galli, E. & Vergari, B. (1985). Earthworm Casting in Plant Propagation. *Horticultural Science 20*, 874-876.

Han, J., Sun, L., Dong, X., Cai, Z., Yang, H., Wang Y. & Song, W. (2005). Characterization of a novel plant growth-promoting bacteria strain *Delftia tsuruhatensis* HR4 both as a diazotroph and a potential bio-control agent against various pathogens. Syst. *Applied Microbiology 28*,66-76.

Hansson, H.G. (1998). NEAT (North east Atlantic Taxa): South Scandivanian marine Annelida. http//www/check-list/southscandivanian/annelids.

Hartenstein, R. &. Bisesi, M.S. (1989). Use of Earthworm Biotechnology for the Management of Effluents from intensively housed livestock. *Outlook on Agriculture*, USA, 18, 72-76

Hendrix, Paul, F., Callaham, Mac A., Drake, John M., Haung, Ching Yu., James, Sam., Snuder, Bruce A. Zhang Weixin (2008). Pandora's box contained bait: The global problem of introduced earthworms. *Annual Review of Ecology, Evolution and Systematics 39*,593-613

Hobbs, R.J. & Harris, J.A. (2000). Restoration ecology: repairing the earth's ecosystems in the new millennium. *Restoration Ecology 9*, 239-246.

Huynh Thi Kim Hoi, Vuong Tan Tu & Nguyen Canh Tien Trinh (2006). Effect of some physic-chemical characteristics of soil on earthworms at Tam Dao National Park. *Vinh Phuc province Khoa* 26: 27-34.

James, S.W., Hong, Y. & Kim, T.H. (2004). New species of *Pheretima* and *Pithemera* (Oligochaeta: Megascolecidae) from Mt. Arayat, Luzon Island, Phiippines. *Revue Suisse De Zooogie 111(1)*,3-10.

Jamieson, B.G.M. (1977). Preliminary descriptions of Indian earthworms (Megascolecidae: Oligochaeta) from the Plani Hills. *Bull. Mus. Natn. Hist. 313*, 477-502.

Julin, E. (1949). De Svenska daggmaskarterna. *Ark. Zool. 43(17)*, 1-58.

Julka, J.M. & Paliwal, R. (1993). Distribution pattern in Indian earthworms In: Earthworm Resources and Vermiculture. *Zoological Survey of India, Solan*, 27-31.

Julka, J.M. & Paliwal, R. (1994). On a new species of *Plutellus* Perrier (Acanthodrilidae: Oligochaeta) from northwest Himalayas, India. *Res. Bull. Punjab. Univ. 44*, 217-220.

Julka, J.M. (1976a). Studies on the earthworms collected during the Daphabum expedition in Arunachal Pradesh, India. *Records of Zoological Survey of India 69*,229-239.

Julka, J.M. (1976b). Studies on the earthworm fauna of Orissa (India); Moniligastridae and Ocnerodrilidae. *Mitt Zool. Mus. Berlin 52*,321-329.

Julka, J.M. (1978). Studies on the earthworm fauna of Orissa (India); Megascolecidae, Octochaetidae and Microchaetidae. *Mitt. Zool. Mus. Berlin 54*,185-197.

Julka, J.M. (1981). Taxonomic studies on the earthworms collected during the subansiri expedition in Arunachal Pradesh, India. *Rec. Zool. Surv. India 26*, 1-37.

Julka, J.M. (1983). A new genus and species of earthworm (Octochaetidae: Oligochaeta) from South India. *Geobios News Reports 2*, 48-50.

Julka, J.M. (1988). The Fauna of India and the adjacent countries. Megadrille: Octochaetidae (Earthworms) Haplotaxida, Lumbricina: Megascolecidae Octochaetidae XIV. *Records of Zoological Survey of India*. 400, 1-20.

Julka, J.M., Blanchart, E. & Chapuis – Lardy, L. (2004). New genera and new species of earthworms (Oligochaeta: Octochaetidae) from Western Ghats, South India. *Zootaxa* 486, 1-27.

Julka, J.M., Giri, S., Panigrahi, P.K. & Senapati, B.K. (1997). *Parryodrilus lavellei* (Octochaetida: Oligochaeta) from Western ghats, South India. *European Journal of Soil Biology 33*,141-144.

Julka, J.M. & Paliwal, R. (2005). Annelida: Oligochaeta in Funa of Western Himalaya. *Records of Zoological Survey of India* 53-90.

Kumar, R., Verma, D., Singh, L.B., Kumar Umesh & Shweta (2010). Composting of sugarcane waste byproducts through treatment with microorganism and subsequent vermicomposting. *Bioresource Technology 101*, 6707-6711.

Launer, J. (2006). The Descent of Man. *OJM. 99*, 275-276.

Lavelle, P. & Spain, A.V. *Soil Ecology*. Kluwer Academic Publiasher, Dordrecht; 2001.

Lee, K.E. (1959). The Earthworm Fauna of New Zealand, New Zealand Department of Scientific and Industrial Research. *Wellington Bulletin 130*,480-486.

Lee, K.E. *Earthworms: their ecology and relationships with soils and land use*. Sydney: Academic press, Sydney; 1985.

Lee, K.E., Blakemore, R.J. & Fraser, P. *The Earthworms of New Zealand*. New Zealand: The Papa Wellington, New Zealand; 2000.

Manum, C.B. & Bose, M.N. Clitellate cocoons: common but neglected fossils amongst Mesozoic plant litter. In: Kielan – Haworowska Z., Heintz, N. &. Nakrem, H.A (Eds.) Fifth Symposium on *Mesozoic Terrestrial Ecosystems: Contributions from the Paleontological Museum*. Norway: University of Oslo, Norway; 1991; 43-44.

Martin, A. (1991). Short and long term effects of endogeic earthworm *Millsonia anomala* (Megascolecidae: Oligochaeta) of tropical savannas on soil organic matter. *Biology and fertility of Soil 11*,234-238.

Mazaud, D. & Bouche, M.B. Introductions sur population et migrations de lombriciens marques. In: Dindal, D.L. (Ed.). *Soil Biology as related to land use practices*. Proceedings of the Seventh International Soil Zool. Colloquium Syracuse. Washington, D.C.: EPA, Washington, D.C.; 1980; 687–701.

Michaelsen, W. (1910). Die Oligochaten fauna der vorde indisch – ceylonischen region. *Abh. Geb. Naturw. Hamburg 19(5)*,1-108.

Michaelsen, W. (1922). Die verbreitung der Oligochaten. *Ver. naturw. ver. Hamb. 29*, 45-79.

Michaelsen, W. (1931). Auslandisshe opisthopore Oligochaeten. Zoologische Jahrbiicher, *Abtheilung Fur Anatomie and Ontogenie der Thiere, Jena. 61*,523-578.

Neilson, R.L. Earthworms and Soil Fertility. USA: Grassland Assoc., New Plymouth, US; 1951.

Neilson, R.L. (1965). Presence of Plant Growth Substances in Earthworms, demonstrated by the Paper Chromatography and Went Pea Test. *Nature 208*: 1113-1114.

Omodea, P. & Rota, E. Taxonomic remarks on the earthworm inhabiting the Western Alps. In: Moreno, A. G. & Borges, S. (Eds.). *Advances in Earthworm Taxonomy*. USA: Editorial complutense, Madrid; 2004; 189-259.

Omodeo, P. (1957). Lumbricidae and Lumbriculidae of Greenland (The Danishs Zoogeographical Investigation of Greenland. Leader: *Christian Vibe). Meddelelser om Gronland 124(6)*, 1-27.

Omodeo, P., Rota, E. & Baha, M. (2003). The megadrile fauna (Annelida: Oligochaeta) of Maghreb. *Pedobiologia 47*, 458-465.

Paliwal, R., & Julka, J.M. (2005). Checklist of earthworms of Western Himalaya, India. *Zoo's print Journal 20*, 1972-1976.

Parle, J.N. (1963). A microbiological study of earthworm casts. *J. of General Microbiology 31*, 13-23.

Patil, B.B. *Soil and Organic Farming*. India: Institute of Natural and Organic Agriculture, Pune, India; 1993.

Perel, T.S. Differences in lumbricid organization connected with ecological properties. In: Lohm, U. & Person, T. (Eds). *Soil Organisms as components of Ecosystems.* Stockholm: Biological Bulletin, Stockholm; 1977; 56-63.

Perrier, E. (1872). Recherches pour Servir à l'historie des Lombriciens terrestres. *Nouv. Arehs. Mus. Hist. Nat. Paris 8*, 5-198.

Piearce, T.G., Oates, K. & Carruthers, W. (1990). A fossil earthworm embryo (Oligochaeta) from beneath a late Bronze Age midden at Potterne, Wiltshire, UK *Journal of Zoology, London 220*, 537-542.

Piearce, T.G., Oates, K. & Carruthers, W.J. (1992). Fossil earthworm cocoons from a Bronze Age site in Wiltshire, England. *Soil Biology and Biochemistry 24 (12)*, 1255-1258.

Pizl, V. (1994). Supplementary records of earthworms (Lumbricidae) in the Czech Republic. *Acta Soc. Zool. Bohem.* 58, 205-211.

Pizl, V. (2002). Earthworms of the Czech Republic. *Sbornik prirodoredeckeho klubu v uherskem Hradisti 9*,151-154.

Plisko, J.D. (1965). Materiay do rezmieszczenia geograficznego I ikologii dzdownic w Polsce (Oligochaeta: Lumbricidae). *Fragm. Faunist 12*, 57-108.

Pop, V.V. & Pop, A.A. (2006). Lumbricid earthworm invasion in the Carpathian mountains and some other cities in Romania. *Invasions Biology 8*, 1219-1222.

Potthoff, M., Joergensen, R.G. & Wolters, V. (2001). Short term effects of earthworm activity and straw amendment on the microbial C and N turnover in a re-moistened arable soil after summer drought. *Soil Biology and Biochemistry 33*, 583-591.

Pramanik, P., Ghosh, G.K., Ghosal, P.K. & Banik, P. (2007). Changes in organic-C, N, P and K and enzyme activities in vermicompost of biodegradable organic wastes under liming and microbial inoculants. *Bioresource Technology 98*, 2484-2494.

Qiu, J.P. & Bouché, M.B. (1998). Revision des taxons supraspécifiques de Lumbricoidea. Doc. Pedozoologiques et Integrologiques. *Montpellier 3*, 179–216.

Reynolds, J.W. (1994). Earthworms of the world. *Global Biodiversity 4*, 11-16.

Reynolds, J. W. & Cook, D.G. *Nomenclatura oligochaetologica. Supplementum primum. A catalogue of names, descriptions and type specimens of the Oligochaeta.* Canada: University of New Brunswick, Fredericton; 1981.

Reynolds, J.W. & Cook, D.G. *Nomenclatura oligochaetologica. Supplementum tertium. A catalogue of names, descriptions and type specimens of the Oligochaeta.* Canada: University of New Brunswick, Fredericton; 1993.

Reynolds, J.W. and Cook, D.G. *Nomenclatura oligochaetologica. A catalogue of names, descriptions and type specimens of the Oligochaeta*. Canada: University of New Brunswick, Fredericton; 1976.

Reynolds, J.W. and Cook, D.G. *Nomenclatura oligochaetologica. Supplementum secundum. A catalogue of names, descriptions and type specimens of the Oligochaeta. New Brunswick Museum Monographic Series (Natural Science).* Canada: University of New Brunswick, Fredericton; 1989.

Rizhiya, E., Bertora, C., Van Vliet, P.C.J., Kuikman, P.J., Faber, J.H. & Van Groenigen, J.W. (2007). Earthworm activity as a determinant for N_2O emission from crop residue. *Soil Biology and Biochemistry 39*, 2058-2069.

Rodríguez, J.A., Zavaleta, E., Sanchez, P. & Gonzalez, H. (2000). The effect of vermicompost on plant nutrition, yield and incidence of root and crown root of Gerbera (*Gerbera jamesonii* H Bolus), *Fitopathologia 35*, 66-79.

Sadhale, Nailini (1996). Recommendation to incorporate earthworms in soil of Pomegranate to obtain high quality fruits. *Asian Agri-History Bulletin,* 1,38-39.

Satchell, J.E. *Earthworm ecology from Darwin to vermiculture*. London: Chapman and Hall Ltd., London; 1983.

Scheu, S. (1989). Microbial activity and nutrient dynamics in earthworms Casts. *J. of Biological Fertility Soils 5*,230-234.

Schwert, D.P. (1979). Description and significance of a fossil earthworm (Oligochaeta: Lumbricidae) cocoon from postglacial sediments in southern Ontario, *Canadian Journal of Zoology 57(7)*, 1402-1405.

Seilacher, A., Bose, P.K. & Pfluger, F. (1998). Triploblastic animals more than 1 billion years ago trace fossil evidence from India. *Science 282*, 80-83.

Senapati, B.K., Julka, J.M., & Dash, M.C. (1990). New species of earthworm *Lennogaster gatu* (Octochaetidae: Oligochaeta) from India. *Rev. Ecol. Biol. Soil. 27*, 467-470.

Sharpley, A.N. & J.K. Syers (1976). Potential role of earthworm casts for phosphorus enrichment of run-off waters. *Soil Biology and Biochemistry 8*, 341-346.

Sims, R.W. & Gerard, B.M. (1985). Earthworms. Keys and notes to the identification and study of the species. *Synopsis of the British Fauna,31*, 171-174

Sims, R.W. (1980). A classification and the distribution of earthworms, suborder Lumbricina (Haplotaxida: Oligochaeta). *Bull Br. Mus. Nat. Hist. Zool. 39*, 103-124.

Sims, R.W. & Gerard, B.M. (1999). Earthworms: notes for the identification of British species. *The Linnean society and the Coastal Science Association. London, UK, 31*: 1-169.

Singleton, D.R., Hendrix, B.F., Coleman D.C. & Whitemann, W.B. (2003). Identification of uncultured bacteria tightly associated with the intestine of the earthworms *Lumricus rubellus. Soil Biology and Biochemistry 35*, 1547-1555.

Speratti, A.B. & Whalen, J.K. (2008). Carbondioxide and nitrous oxide fluxes from soil as influence by anecic and endogeic earthworms. *Applied Soil Ecology 38*, 27-33.

Stephenson, J. (1914). On a collection of Oligochaeta mainly from Northern India. *Records of Indian Museum 10*, 321-365.

Stephenson, J. (1915). On some Oligocheata, mainly from Southern Indian and Ceylon. *Records of Indian Museum 6,*35-108.

Stephensosn, J. (1916). On a collection of Oligocheata belonging to the Indian Museum. *Records of Indian Museum 12*,299-354.

Stephenson, J. (1917). On a collection of Oligocheata from various parts of India and further India. *Records of Indian Museum 13*,353-416

Stephenson, J. (1922). Some earthworms from Kashmir, Bombay and other parts of India. *Records of Indian Museum 24*, 427-443.

Stephenson, J. *Oligochoeta: The Fauna of British India, including Ceylon and Burma.* London: Taylor and Francis Limited, London; 1923.

Stephenson, J. *The Oligochaeta.* U.S.A: Oxford University Press, U.S.A.:1930.

Suhane, Ravindra K., Sinha, Rajiv K. & Singh, Pancham K. *Vermicompost, cattle-dung compost and chemical fertilizers: impacts on yield of wheat crops.* India: Rajendra Agriculture University, Pusa, Bihar, India; 2008.

Sutherst, R.W. (2000). Climate change and invasive species: a conceptual framework. Island press, Washington. *Swil and Hund 90(11)*, 14-19.

Szczech, M., Rondomanski, W., Brzeski, M.W., Smolinska, U. & Kotowski, J.F. (1993). Suppressive effect of a commercial earthworm compost on some root infecting pathogens of cabbage and tomato. *Biological Agriculture and Horticulture 10*, 47-52.

Templeton, R. (1844). Description of *Megascolex caeruleus. Proceedings of Zoological Society London 12*, 89-91.

Teotia, S.P., Duley, F.L. & McCalla, T.M. (1950). Effect of stubble mulching on number and activity of earthworms. *Agricultural Experiment Research Station Bulletin 1*,160-165.

Terhivuo, J. (1988). The Finnish Lumbricidae (Oligochaeta) fauna and its formation. *Annales Zoologici Fennici 25*,229-247.

Terhivuo, J. (1989). The Lumbricidae (Oligochaeta) of southern Finland: species assemblages, numbers, biomass and respiration. *Annales Zoologici Fennici 26*,1-23.

Thai, Tran Bai & Do, Van Nhuong. (1989). Remarks on earthworm fauna in Phnom Phenh and the neighboring regions. *Thong Bao Khoa Hoc 1,* 76-79.

Thai, Tran Bai & Samphon, S. (1991). List of discovered species in Laos. *Truongdaihocsupham Hanoi 1*,86-89.

Thai, Tran Bai. (2000). Da Dang Loai Giun Dat O Viet Nam. Species Diversity of earthworms in Vietnam. *Nung van De Nghien Cun co Ban Trong Sihn Hoc 1,* 307-311.

Tiunov, A.V., Hale, C.M., Holdsworth & A.R., Vsevolodova–Perel, T.S. (2006). Invasion patterns of Lumbricidae in to the previously earthworm–free areas of northeastern Europe and the Western Great Lakes region of North America. *Biological invasions 3*,1223-1234.

Tomati, V. & E. Galli (1995). Earthworms: Soil fertility and plant productivity. *Acta Zoologica Fennica 196*,11-14.

Tomati, V., Grappelli, A. & Galli, E. Fertility factors in earthworm humus. In: *Proc. of International Symposium on Agriculture and Environment: Prospects in Earthworm Farming*; Rome; 1983;49-56.

Tomati, V., Grappelli, A. & Galli, E. The presence of growth regulators in earthworm-worked wastes. In: *Proceeding of International Symposium on Earthworms*; Italy; 1987: 423-436.

Tomati, V., Grappelli, A. & Galli, E. (1988). The hormone-like effect of earthworm casts on plant growth. *Biology and Fertility of Soils 5*, 288-294.

Trigo, D., Mascato, R., Iglesias Briones, M.J. & Diaz Cosin, D.J. (1990). Lombrices de tierra de Portugal continental. Inventario y citas. *Arquivos do Museu Bocage Nova Serie 1(38)*, 521-567.

Tsai, C.F., Shen, H.P. & Tsai, S.C. (2000). Native and exotic species of terrestrial earthworms (Oligochaeta) in Taiwan with reference to northeast *Asia Zool. Stud. 39(4)*, 285-294.

Tsai, C.F., Shen, H.P. & Tsai, S.C. (2004). Endemicity and altitudinal stratification in distribution of megascolecid earthworms in the centro western Taiwan. *Endemic species Research 6(2)*,1-18.

Valentine, J.W. (1980). L'origine des grands groupes d'animaux. *La Recherche. 112*, 666-674.

Verma, D. & Shweta (2011). Earthworm resources of western Himalayan region, India. *Int. J. Soil Sci. 6*,124-133.

Verma, D., Bharti, S. & Yadav, S. (2010). Earthworm resources in Gangetic plains of India. *International Journal of Biodiversity and Conservation 2*,134-139.

Wallwork, J.A. *Earthworm Biology*. India: Gulab Vazirani for Arnold-Heinemann, New Delhi, India;1984.

Wegener, Alfred L.1996. German meteorologist, Arctic explorer and proponent of the theory of Continental Drift (1915, 1924), the forerunner of modern plate tectonics. http://onlinelibrary.wiley.com/doi/ 10.1002 /9780470750643.app1/pdf.

Whalen, J.K., Robert Parmelee, W., David McCartnney, A., Vanarsdale, Jessica L. (1999). Movement of Nitrogen (N) from decomposing earthworm tissue to soil, microbial and nitrogen pools. *Soil Biology and Biochemistry 31*, 487-492.

Zaller, J.G. (2006). Foliar spraying of vermicompost extracts: Effects on fruit quality and indications of late-blight suppression of field-grown tomatoes. *Biological Agriculture and Horticulture 24*, 165-180.

Zicsi, A. & Csuzdi, C.S. (1997). Uberweitere regenuramer aus Ekuador. Regenurimer aus Sudamerika. 28. *Berichte des naturwissen chaftlichmedizinischen. Vereins in Innsburck 84*, 81-103.

Zicsi, A. & Michalis, K. (1981). Übersicht der Regenwurmer Fauna Griechenlands (Oligochaeta: Lumbricidae). *Acta Zool. Hung. 27 (1-2)*, 239-264.

Zicsi, A. & Csuzdi, C. (1999). Weitere Angaben Zur regenwurm-Fauna Frankreichs Sowie die Beschreibung fung newer Arten (Oligochaeta:

Lumbricidae. (Further contribution to the earthworm fauna (Oligochaeta: Lumbricidae) of France, with description of five new species and one subspecies). *Revue Suisse de Zoologie 106(4)*, 983-1003.

Zicsi, A. (1981). Wietere Angaben zur Lumbriciden fauna Italiens (Oligochaeta: Lumbricidae). *Opusc. Zool. Budapest 17/18*,157-180.

Annex I: Organic Input Producers for Fertility Management in India

1. Agriculture Tools Research Centre, Suruchi Compus, Bardoli-394601, Gujarat .
2. Amrit Biotech, Commercial Vermiculture production, Pinjarapole Gaushala, Sanganer, Jaipur-302001, Rajasthan.
3. Anand Bhai Patel Farm, Demonstration and Commercial Vermiculture Production, Khirai Paper, Bhuj-370001, Gujarat.
4. Asia Pacific Natural Agriculture Network, 35 Gandhi Road, Dehradun-248001, Uttaranchal.
5. BAIF Development Research Foundation (BAIF), Pradeep Chamber, Bhandarkar Road, Pune- 411004, Maharasthtra.
6. Bhavalkar Earthworm Research Institute, A/3, Kalyani, Pune-411037, Maharashtra.
7. Bhilwara District Milk Union (Vermicomposting CFC), Milk Union Processing Plant, 5 Kms Ajmer Road, Bhilwara-311001, Rajasthan.
8. Central Sericultural Research and Training Institute, Research organization for utilization of Sericulture waste or Vermiculture vermicompost, Central Silk Board Srirampura, Mysore-570008, Karnataka.
9. Diamond Agrotech and Biofertilizers, Diamond House, More Hospital Road, Navalgarh-333042, Rajasthan.
10. Gandhigram Trust, Gandhigram-624302, Tamil Nadu.
11. Go-Vigyan Research Centre, Nagpur, Kamdhenu Bhavan, Mahal, Nagpur-440002, Maharashtra.
12. Indian Social Institute, Literature and Research for Organic Manures, New Delhi.

13. Institute of Natural Organic Agriculture (INORA), Research and Developments in Vermiculture, Erandwan, Pune-411004, Maharashtra.
14. Institute of Natural Organic Development Trust, 100 Kalpana Apartments Prabhat Road, Erandavana, Pune- 411004, Maharashtra.
15. Institute of Organic Farming and Technology, B-9, Shivai Bank of Maharashtra Colony, Sinhagad Road, Pune-411030, Maharashtra.
16. Institute of Research in Soil Biology and Biotechnology, c/o Dr. Sultan Ismail, The New College, Chennai-600014, Tamil Nadu.
17. Khadi and Village Industries Commission (KVIC), Promotional and Development Agency for Bio-Manures, Gramodaya, 3 Irla Road, Ville Park (W), Mumbai-400056, Maharashtra.
18. M.S. Swaminathan Research Foundation, Vermiculture and Vermicompost, Tharamani Institutional Area, Chennai-600014, Tamil Nadu.
19. Parry Agro Industries Ltd., Tiam House Annex 2 Jehangir Street, Joint Initiative for Organic manure – F.B.O. Technology Development, Chennai-600001, Tamil Nadu.
20. Patil Vermi Farm, R.S. Patil “Shiva” Behind Prasad Lodge, Gadag-582101, Kanataka.
21. R.P.G. Life Sciences Ltd., 4th Milestone, Sirsa Road, Hisar-125001, Haryana.
22. Raghovar Agrotech Industries, Sunjeevan Farms. Beawar Road, Pisangan, Ajmer-305001, Rajasthan.
23. Rashtriya Organic Fertilizer, D/97 Ambabari Jaipur-302012, Rajasthan.
24. Ramabh Bhai Organic Farm, Commercial Vermiculture, Production on Organic Papaya Farm Kukama, Bhuj -370001, Gujarat.
25. Supa Biotech Pvt. Ltd., Vermiculture, Liquid Fertilizer, Biodynamic Preparation, Kohinoor Building Dina Lodge Compound, Mallital, Nanital-263001, Uttarakhand.
26. T & I Global Ltd., Commercial Vermiculture Unit Tea Estate under conservation for Organic farming, 19-RN, Mukherji Road, Calcutta-700001, West Bengal.
27. The Energy Research Institute (TERI), Darbari Seth Block, Habitat Place, Lodhi Road, New Delhi-110003.
28. Texas organics Pvt. Ltd., Micro Nutrient Preparations, M.C. Road, Pandalam, Pathanamthitta-691523, Kerala.

29. Trees for Life Green World Associates, Household, Conservation by Vermiculture, Flat No. 620, Sector 19, Pocket-II, Dwarka, New Delhi-110045.
30. U.P. Diversified Agriculture Support Project, Development and Promotional Vermicompost, Vermiwash, Block B/B2 IV Floor, P.I.C. U.P. Bhavan, Gomati Nagar, Lucknow-226001, Uttar Pradesh.
31. Vishva Bharti Kisan Jagriti Shikshan Sansthan, Development and Promotion of Natural Fertility management, Shree Ladwa Goshala, Ladia, Hisar-125087, Haryana.
32. Yashvati Earthworms and Pest Research Institute, Tulsiram Road, Tinsukiya -786125, West Bengal.

ANNEX II: DIRECTORY OF VERMICULTURE RESOURCES

A & J Worm Farm, Arnold Janke Loop, Elm Mott Tx 76640, USA.

A&A Worms, Mike Laskey, 32804 Texas Drive, Yucaipa, CA 92399, USA, E-mailfishingworms@prodigy.net, http://www.a-aworms.com.

ACL Bio Products Limited, 6 Josier Street, Nungamabakkam, Chennai-600034, India.

Advanced Biotechnology Inc., Bill Kreitzer (Sells encapsulated earthworm eggs/cocoons, Called ET Seeds for soil application), 104 W. Market St. P.O. Box 1, Elliott. IL 60933, USA.

Agricare Organic Farms, 8, Ff, Madhok Trade Centre, Ludhiana -140001, Punjab, India.

Agriculture Tools Research Centre, Suruchi Campus, Bardoli-394601, Gujarat, India.

Albrecht & sons Work Farms., Buzy Albrecht, Kerman, California, USA, E-mail: klalbrecht@uswest.com.

Ambrose Worm Products, RD.2 Box 401A, New Florence, P.A. 15944, USA, www.http:homepage.Third_wave.com/wormpage/worms, vermicompost.

American Resource Recovery, Mario Travolini, P.O. Box 315, Westley CA 95387, USA.

AmerigrowFarms, Carlos Leanne Herrera, USA, Email: clay.Herrera@ prodigy.net.

Amrit Biotech, Pinjarapole Gaushala, Sanganer, Jaipur-302001, Rajasthan, India.

Amway India Enterprises, A-5, Kailash Colony, New Delhi-110048, India.

Anand Bhai Patel Farm, Khirai, Rapar, Bhuj-370001, Gujarat, India.

Anderson Worms limited, Tim Anderson, 11270 Arno Road, Galt, CA 95632, USA.

Annelids and More, Bob and Sherin Bowen, P.O. Box 301, Stevens Point, WI 54481, USA, E-mail: rbowen@coredcs.com.

Appropriate Technology Transfer for Rural Areas (ATTRA), Alice Beetz, Distributes appropriate technology information, P.O. Box 3657, Fayetteville, AR 72702, USA.

ARBICO, P.O. Box 4247, 18701 N. Lago Del oro Parkway, Tucson; AZ 85739, USA, E-mail: ARBICO@aol.com.

Archives on the Internet, Lawrence London, Venaura Farm, 7119 Old Greensboro Road, Chapel Hill, NC 27516, USA, E-mail: London@ sunSITE.unc.edu, www.sunsite.unc.edu:70/11/pub/academic/ agriculture/sustainable_agriculture/vermiculture.

Arkanasas Wigglers Worm Farm, Michael Ford, 532 Bunker Road, Harrison, AR 72601, USA, http://www.wso.net/arkansaswigglers.

Arland & Sons Innovative Gardening Products, 11881 Arroyo Avenue, Santa Ana, CA 92705, USA, E-mail: arlan99@yahoo.com, http://www. arlanson.com.

As The Worm Turn, Tina Creemmett, 5020 Christofferson Road, Turlock. CA. 95380, USA, E-mail: mctina320@aol.com.

As the worm turns, Mitch & Libby Vassar, 2765 Sunset Road, Bishop CA. 93514, USA.

Asia Pacific Natural Agriculture Network 35, Gandhi Road, Dehradun-248001, Uttaranchal, India.

Avant Garden Vermicomposting Systems, P.O. Box 1047, Point Reyes Station, CA 94956, USA.

Baif Devlopment Research Foundation (BAIF), Pradeep Chambers, Bhandarkar Road, Pune-411004, Maharashtra, India.

Bassett's Cricket Ranch, Inc, Russell Bassett, S. Mariposa, Visalia, CA 93292, USA, http://www.bccricket.com/.

Bate's Baite, George.Bates, 338 Bruce Costin Road, Warsaw, NC 28398, USA.

Beaver River Associates, Joshua Nelson Vermont, USA, E-mail: RIWiggler@aol.com, http://MerchantsBay.com/BeaverRiver.

Beeson's Worm Farm, Richard C. Beeson, 1826 E. Elmore, Crawfordsville, IN 47933, USA.

Berkeley Worms, P.O. Box 4126, Berkeley, CA 94704-0126, USA, E-mail: compost@ocf.berkeley.edu, http://www.ocf.berkeley.edu/%7Ecompost/ index.html.

Bhavalkar Earthworm Research Institute A/3, Kalyani, Pune-411037, Maharashtra, India.

Bhawalkar's Earthworm Research Institute, UdayBhawalkar, Vermiculture biotechnology and farm-scale vermicomposting, A/3 Kalyani, Pune-411037, India.

Bhilwara District Milk Union Milk Union Processing Plant, 5 Km Ajmer Road, Bhilwara-311001, Rajasthan, India.

Bio Control Research Laboratories (PCI), P.O. 3228, 479, R.N. Nagar, Bangalore-560032, Karnataka, India.

Bio Era Technologies Pvt. Ltd., B-15, Cropn. Buld, Link Road, Nagpur-440001, Maharashtra, India.

Bio State Agri Sciences, A Division of Wockhardt Life Sciences Ltd., Wockhardt Towers, Bandra-Kurla Complex, Bandra, Mumbai-400059, Maharashtra, India.

Biogro, Bv. Artigas, 1919/203 Montevideo, Uruguay, bioagro@multi.com.uy, http://webs.montevideo.com.uybioagro.

Biological Home Grown Farms, Tom Bennington, P.O. Box 3458, Riverside, CA 92519-3458, USA, E-mail: biohome@pacbell.net, http://www.worm-publications.com/biologicalhome.htm.

Biotech International Ltd., 35 A Vishal Tower, Indra Complex, Navilakha, Indore-452001, Madhya Pradesh, India.

Blue Belly Farm, Todd Spratt, Vermitech specialist, 20244 Poker Flat Road, Penn Valley, CA 95946, E-mail: info@bluebellyfarm.com, http://www.bluebellyfarm.com.

Blue Ridge Redworms, James G. Magee, 58 Old Farm School Road, Asheville NC 28805, USA.

Bond Worm Farm, Inc., Wanda Bond, 900 Bliss Road, Ceres, CA 95307, USA.

Boot's Worm Farm, George Parker, Rt. 7 Box 574-A, Lumberton NC 28358, USA.

Borderlines, Inc, Steve Rear, 1602 Filipov Drive, Champaign, IL 61821, USA, E-mail: Borderlines@email.msn.com.

Bountiful Gardens, 18001 Shafer Ranch Road, Willits, CA 95490-9626, USA, E-mail: bountiful@zapcom.net.

Brookside Worm Farm, Bob Thomson & Ivan Broyles, P.O. Box 386, Gowanda, NY 14070-0386, USA.

Brown's Worm Farm, Rosemary Brown, P.O. Box 284, Marietta, OK 73448, USA.

Bruce's Worms-4-All, Bruce Hawkins, 37 Jacaranda Avenue, Hollywell 4216 Queensland, Australia http://www.infonet.net.au/worms/.

Bucks County Worm Farm, Dave & Teresa Greninger, 5125 Route 313 Swamp Road, Mountainville, PA 18923, USA, E-mail: bcwf@aol.com.

Burke's Worm Farm, Jim Burke, 651 Shadyglen Drive, Baton Rouge, LA 70816, USA, E-mail: wormgrower@hotmail.com.

C & C Vermiculture, Charmaine Harris & Carolyn Foxe, 527 N, Shirk Road, Visalia, CA 93291, USA, E-mail: CCVERM@hotmail.com.

C & G Enterprises, Carol Miller, P.O. Box 658, Perry, OH 44081, USA.

CADILA Pharaceuticals Ltd. (Agro Division), 'CADILA Corporate Campus' Sarkhej-Dholka Road, Ahmedabad-382210, Gujarat, India.

California Worm Growers, Mike Chamberlain, 1308B North Inyo Street, Ridgecrest, CA 93555, USA, E-mail: Bob&Gin@ridgenet.net; cal_worm_growers@hotmail.com.

Cape Cod Worm Farm, Maggie Pipkins, 30 Center Avenue, buzzards Bay, MA 02532, USA, E-mail:Capeworms@aol.com, http://members.aol.com/ Capeworms/private/whrm home. htm.

Carey's Worm Ranch, Mark Carey, 106 Pioneer Drive, Minooka, III, USA, Carolina Worm Farm, Barbara Corson, 209 Calhoun Street, Clover, South Caroina 29710, USA.

Cascade Vermicomposting System, Will Gehr, 3431 Knolbrook Avenue, Corvallis, OR 97333, USA, E-mail: willgehr@hotmail.com.

Casting a New Future, Wendy Anderson, 3635 SW Custer Street, Portland, OR 97219, USA.

Center for Agroecology & Sustainable Food systems, Dr. Mathew Werner, University of Caifornia, 1156 High Street, Santa Cruz, CA 95064, USA, E-mail: Werner@zzyx.ucsc.edu.

Central Sericulture Research and Training Institute Central Silk Board, Srirampura, Mysore-570008, Karnataka, India.

Centre for Indian Knowledge Systems, Gandhi Mandapam Road, Kotturpuram, Chennai-600085, Tamil Nadu, India.

Chitra Exports, 572, Mettupalyam Road, North Coimbatore, Coimbatore, Tamilnadu, India, chitraherb@yaoo.co.in; achitra@vsnl.net.

Chuck's Rod'n Reel, Springfield, OR, USA.

City Worms & compost, 1850 32nd Avenue, San Francisco, CA 94112, USA.

Composters.com, Karl P. Warkomski, Laguna Beach, CA 94112, USA, E-mail: kpw1@greenculture.com, http://www.composters.com.

Connecticut Valley Worm Farm, Jim Dombrowski, P.O. Box 206, Enfield, CT 06082, USA, E-mail: connvalley@aol.com.

Cosmos's Red Worms, Paul Cosmides, 432 Lawton Street, San Francisco, CA 94122, USA, http://www.alcasoft.com/cosmos.

Cravens Worm Farm, Darrel & Linda Cravens, Rt. 6, Box 174, Liberty, KY 42539, USA.

Creative Composting Concepts, Stacie Johnson, 500 W, Main Street, Robins, IA 52338, USA, E-mail: soilcycle@aol.com, http://www. wormswork.com.

Creekside Gardens, Sherry Palmer, 155 Nix Road, Chehalis, WA 98432, USA E-mail: creekside@wormlady.com, http://www. wormlady.com.

Crow Worm Farm, Bobby & Betty Crow, 4808 FM 2135, Cleburne, TX 76031, USA.

CSIRO, Davies Laboratory, Dr. A.V. Spain, Private Mail Bag, Aitkenvale, Queensland 4814 Australia.

Cyclone Manufacturing Company, 8697 Island Road, Waterloo, WI 53594, USA.

D & D Ranch, Deb and Dave Hoffman, 1450 Quirrel Ridge Road, NW, Swisher, IA 52338, USA, E-mail: Deb_Hoffman@mail.ccs.k12.ia.us.

Dakota Blue Worms, 432 Swift Creek Road, Cordele, GA 31015, USA, E-mail: paris@dakotablue.com.

Danny & Wanda's Worm Farm, 2730 Crooked Creek Trail, Boonville, NC 27011, USA, E-mail: worms@rivercto.net.

Darwin's Farm, Luciane Berg, P.O. Box 8010, Cedar City, UT 84720, USA, E-mail: Darwinsfarm@infowest.com;Luciane000@hotmail.com.

DeArmond's worm Farm, Don & Sherri DeArmond, 3923 Greenbck Road, Utica, KY 42376, USA.

Diamond Agrotech and Bio Fertilizers Diamond House, More Hospital Road, Navalgarh-333042, India.

Down 'n Dirty Farms, Jenise Miller, 1030 County Road 176, Gatesville, TX 76528, USA.

dpi-design, inc., Box 10, Site 10, RR1, Thorsby, AB (Alberta), T0C 2P0 Canada, E-mail: dpi@telusplanet.net, http://www.doityourselfkit.com.

Dusty Morgan's Worm Farm, Jordana Adler, 1336 Moorpark Road, # 127, Thousand Oaks, CA 91360, USA, E-mail: jnjorge@gte.net.

E.I.D. Parry Ltd., "Dare House", 234, NSC Bose Road, Chennai-600001, Tamil Nadu, India.

Early Bird Worm Farm, Paul & Lori Coleman, 6902 Hwy 25 North, Hodges, SC 29653, USA,E-mail: earlybirdworms@yahoo.com, http://www. earlybirdwormfarm.com.

Earth Angel Worm & Garden, WORM; 888-BUY WORMS, (Worms, supplies, literature, consultation), Karin Kelley, USA, E-mail: earthangel@ buyworms.com.

Earthly Good Products, Ken Williams (Vermiculturalist specializing in rural applications from waste reduction to soil improvement), 232 Cascade Street, Raceview, 4305, Southeast Queensland, Australia, E-mail: egprods@gil.com.au.

Earthworm Institute, Larry Wilhelm (Vermicomposting resources), P.O. Box 95104, Grapevine, TX 76099, USA.

Earthworms & Supplies, Dave Metzger, 2144 Hog Trough Road, Owensville, MO 65066-9549, Canada.

Earthworms for Ecology, Michael & Natalie Benavidez, P.O. Box 6473, Sante Fe, NM 87502, Canada, E-mail: worm1man@aol.com, http://flash.net/ ~phidel/redworms.html.

Ecoway Environmental Products Distribution, Mitchell and Myriam Goldberg, 3171 Samantha Drive, Santa Ynez, CA 93460, USA, E-mail: ecoway@silcom.com, http://www.wp.com/51316/.

Enviro Care of America, Worm Wigwams, starter kits, Wyoming, USA, E-mail: info@envirocare.net.

Enviro-Ganics Ecological Systems, Inc, Sandra Kandracs, P.O. Box 20241, Grantham Postal Station, St. Catherines, Ontario L2M 7W7, Canada.

Environmental Earthworm Products, Inc., Frank Stevenson, 8114 Port Said Street, Orlando, FL 32813, USA.

Environmental Recycling Systems, Large-scale vermicomposting systems, P.O. Box 904, Alpine, CA 91903, USA.

Environmental Science Policy Management, Dr. Paul Gersper, Vermiculture research in Cuba, 108 Hilgard Hall, 3110, University of California at Berkeley, Berkeley, CA 94720-3110, USA, E-mail: gersper@ nature.berkeley.edu.

EPM, Inc.; Bruce Elliott, On-site food scrap composting products, consulting, and sales, P.O. Box 1295, Cottage Grove, OR 97424, USAE-mail: bruce@fish.world.com, http://www.wormwigwam.com.

Farm Services Indore Group 6, Sikh Mohalla, Main Road, Indore-452007, Madhya Pradesh, India.

Flowerfield Enterprises, Mary Appelhof (Worm Woman), 10332 Shaver Road, Kalamazoo, MI49002, USA, E-mail: mary@wormwoman.com, http://www.wormwoman.com.

Foley Farms, Pat Foley, P.O. Box 617, Coarsegold, CA 93614-0617, USA, E-mail: foleys@sierratel.com/foleys/.

Fred's worm Ranch, Fred Patz, 1409 N. 60th, Lincoln, NE 68505, USA.

Gandhigram Trust Dindigul, Gandhigram-624302, Tamil Nadu, India.

Gardener's Supply Co., 128 Intervale Road, Burlington, VT 05401, USA, E-mail: info@gardeners.com, http://www.gardeners.com.

George Morris, 6043, Fraiser Street, Georgetown, SC 29440, USA.

Gess Environmental, Andrew Jolin, 4591 Kjer Road, McKinleyville, CA 95519, USA, E-mail: gesenv@aol.com.

Global worming, Bruce Hamilton, P.O. Box 745, Inverness, CA 94937, USA.

Good Earth Company, 702 North 1st Avenue, Suite C, Yakima, WA 98902, USA, E-mail:sales@wormfarm.com, http://www.wormfarm.com.

Go-Vigyan Research Centre, Nagpur Kamdhenu Bhavan, Mahal, Nagpur-440002, Maharashtra, India.

Grace Exports, No. 3, Sivanthipatti Road, Opp To Pushpalatha School, Tirunelveli, Tamil Nadu -627011, India.

Griffin Farms, Ed & Pat Griffin, 8401 Crossroads, Benton, Ar 72015, USA.

Gujarat state Forest Development Corporation Ltd., Bunglow No. 198/4, GH-1, Opp. Civil Court, Sector-30, Ghandinagar-382010, Gujarat, India.

H & K Research and Development Co., Inc., P.O. Box 44184, Omaha, NE 68144, USA.

Hall's Organic Gardens, Ezell Hall, 1805 Colvert Street, Tuskegee Institute, AL 36088, USA.

Happy chick Worm Ranch, Howard & Marie Mays, 14180 Mays Rd., Gulfport, MS 39503, USA.

Happy D Ranch Worm Farm, Al & Dorothy Benoy, P.O. Box 3001, Visalia, CA 93278, USA, info@happydranch.com, http://www.happydranch.com.

Harmony Farm supply, 3244 Gravenstein Highway, North Sebastopol, CA 95472, USA, Phone: (707) 823-9125, Email:info@harmonyfarm.com, http://www.harmonyfarm.com.

Herban Renewal Herb Garden & Nursery, 10437 19th Avenue SW, Seattle, WA, 98146, USA.

Iannone Enterprises, International, Marcia Iannone, 8469 18th Street, Alta Loma, CA 91701, USA, miannone@earthlink.net.

Indian Institute of Soil Sciences Indore-452001, M.P., India.

Indian Social Institute, New Delhi Institute of Natural Organic Agriculture (INORA) Erandwan, Pune-411004, Maharashtra, India.

Indore Biotech Inputs and Research Pvt. Ltd., 6, Sikh Mohalla Main Road, Indore-452007, Madhya Pradesh, India.

Institute of Natural Organic Agriculture (INORA), Professor M.R. Bhiday, 100, Kalpana Apartments, Erandwan, Pune 411004, Maharashtra, India.

Institute of Natural Organic Development Trust 100, Kalpana Apartments, Prabhat Road, Erandavana, Pune-411004, Maharashtra, India.

Institute of Organic Farming and Rural Technologies B-9 Shivai, Bank of Maharashtra Colony, Sinhagad Road, Pune-411030, Maharashtra, India.

Institute of Research in Soil Biology and Biotechnology The New College, Chennai-600014, Tamil Nadu, India.

Interior Gardens, 2817 Lyndale Avenue South, Minneapolis, MN 55408, USA, E-mail: interiorgardens@interiorgardens.com.

International Panacca Limited, 11, Babar Lane, Bengali Market, New Delhi-110001, India.

Isaac McGill, 108 King Avenue, Kingstree, SC 29556, USA.

J & S Worm Farms, Jay & Scott Downs, 411S, 4th Avenue, Paragould, AR 72450, USA.

J & T's Redworms, Tim Matthews, Blue Lake, CA, USA, dairyworms@humboldt1.com, http://www.humboldt1.com/dairyworms.

Jabour's Worm Farm, 33050 Stagecoach Road, Nuevo, CA 92567, USA, E-mail: mjworm@pe.net.

James and Edna Cooper, Route 1, Box 110-B, Nesmith, SC 29580, USA.

Jayam Bio Manures Pvt. Ltd., Chennai, 14/2 (Old No. 38/2), Akbarabad Ii Street, Kodambakkam, Chennai, Tamil Nadu -600024, India.

Jenkins Creek Farms, Rich and Susan Beasley, P.O. Box 1482, Flippin, AR 72634, USA, elaphe@southshore.com, http://www.jenkinscreek.com.

Jet Worm Farm, Tracy Laster, P.O. Box 7736, Jackson, TN 38308, USA.

Jim Heberle, 3815 Route 96, RD #2, Shortsville, NY 14548, USA.

Jim Knight, Rt. 2, Box 183, Eudora, AR 71640, USA, E-mail: jimbo@seark.net.

JS International, 74, Pushpanjali, Vikas Marg, Delhi-110092, India.

JWH Industries, Michael James, P.O. Box 1285, Florence, SC 29503, USA.

K Anne W Farms, Gary Williamson, 6344 County Home Road, Winterville, NC 28590, USA.

Kamal Nursery, Andul-Mouri, Howrah-711302, Uttar Pradesh, India.

Kanoria Chemicals & Industries Limited, Park Plaza, 71, Park Street, Kolkata-700016, West Bengal, India.

Kanpur Gaushala Society, 55/112 Gaushala Bhavan, General Ganj, Kanpur-208001, India.

Kazarie Worm Farm, Dan Warco, 7370 SE 56th Terrace, Trenton, FL 32693, USA, E-mail:kazarie@afn.org, http://www.kazarie.com.

Ken's Bait & Tackle, Maplewood Plaza, 2612 Union Avenue, San Jose, CA 95130, USA.

Khadi and Village Industries Commission (KVIC), Gramodaya, 3 Irla Road, Ville Parle (W), Mumbai-400056, Maharashtra, India.

L & T Enterprise Worm Ranch, Lyle Dunham, 119 North 8th, Humboldt, KS 66748, USA.

Lake County Worm Farm, Richard Morhar, P.O. Box 1332, Kelseyville, CA 95451, USA, E-mail:wormfarm@pacific.net, http://www.pacific.net/~wormfarm/.

Larry Judkins Worm Farm, Larry Judkins, 29172 Ranchito Road, Helendale, CA 92342, USA.

Laurie and Inetta Cooper, Route 1, Box 512, Hemingway, SC 29554, USA, E-mail: SeaJay720@aol.com.

Lazy G Bar Farm, Keith Grimes, P.O. Box 70276, Montgomery, AL 36107, USA.

Leach's Bait Farm, Jack Leach, Rt. 1, Box 296, East Bernstadt, KY 40729, USA.

Lil Wiggler Farms, Robin Looney, Rt. 1, Box 59-2, Coweta, OK 74429, USA, E-mail: looneyfarm@peoplepc.com.

Lost Mountain Environmental Enterprises, Inc., R.L. Jacobs, P.O. Box 1175, Marietta, GA 30061-1175, USA,E-mail: rljr44@aol.com, http//: home.mindspring.com/~jjacobos002/homepage.htm.

Lucky Hooker Worm Farm, John & Velma Warm, 295N, State Street, Ferron, UT 84523, USA.

M S Swaminathan Research Fondation, Tharamani Institutional Area, Madras-600113, India.

M S Swaminathan Research Foundation Tharamani Institutional Area, Madras-600113, Tamil Nadu, India.

M. R. Morarka GDC Rural Research Foundation, Station Road, Nawalgarh - 333 042, Rajasthan, India.

M.R. Morarka Research Foundation, Ward No.16,16, Siwal Colony, Jaipur - 303901, Rajasthan, India, E-mail: info@morakango.com.

Mac's Red Worm Farm, Isaac McGill & Sons, 108 King Avenue, Kingstree, SC 29556, USA.

Magic Wiggler Worm Ranch, Larry Pelletier, Owner, Marie Bateman, Business Manager, 1130 Bush Road, Raymond, MS 39154, USA, E-mail: sales@magicwiggler.com, http://www.magicwiggler.com.

Magic Worm Ranch, Karl Langer, 3163 Roadrunner Road, San Marcos, 877-Worm-Boy or (760) 598-0607, CA 92069, USA, E-mail: Karl@magicworms.com.

Maharashtra Agricultural Bioteks, Dr. Hemangree Jambhekar, B/9 Shivai Housing Society, Near Sane Guruji Smarak, Pune- 411030, Mharasthra, India.

Manchester Worm Farm, James Plourde, 1131-0 Tolland Tpke, Manchester, CT 06040, USA;info@marchbiological.com, http://www.marchbiological.com.

Compound, Angeripalyam, Gani Nagar, Tiruppur-641603, Tamil Nadu, India.

Mel's Worm Farm, Arland Malmstrom, 450 S. 100 East, Price, UT 84501, USA.

Melton Bait Farm, Butch Melton, P.O. Box 895, Manning, SC 29102, USA.

Mike Moffatt, 3833 NE 4th Street, Ocala, FL 34470, USA.

Monterey Worm Farm, Preston Welch, 9518 Hwy 129, Monterey, LA 71354, USA.

Morning Mist Worm Farm, James & Karen Cain, P.O. Box 1155, Davis, California 95617, USA, E-mail: mmcain@communityonline.net, http://morningmistwormfarms.com.

Mose and Stanley Cooper, 706 Benson Street, Kingstree, SC 29580, USA.

Mountain Creek Worm Farms, Tim Smith, 40779 Hearne Road, New London, NC 28127, USA.

Mountain High Compost, Santa Cruz, CA, USA.

Multiplex Bio-tech Pvt. Ltd., 180, 1st Main Road, Mahalakshmi Layout Extension, Bangalore-560001, Karnataka, India.

National Centre for Integrated Pest Management, ICAR, Lal Bahadur Shastri Bhavan, Pusa Complex, New Delhi-110012, India.

Natural Gardening Co., David Baldwin, 217 San Anselmo Avenue, San Anselmo, CA 94960, USA, E-mail: info@naturalgardening.com, http://www.naturalgardening.com/shopping/.

Natural Life Science Pvt. Ltd., 1st Floor, Madhav Apts., Shridhar Nagar, Chinchwadgaon, Pune-411033, Maharashtra, India.

Navdanya Research Foundation, A 60 Hauz Khas, New Delhi-110016, India.

Nicholson's Worms, John F. Nicholson, 873 Green Acres Drive, Hernando, MS 38632, USA, E-mail: GardenersGold@juno.com.

NIIR Project Consultancy Service, 106-E, Kamla Nagar, Delhi-110007, India.

Nixon & Sons Live Baits, Mike Nixon, Rt. 2, Box 63, Evington, VA 24550, USA.

North Carolina State University, Rhonda Sherman, Biological & Agricultural Engineering Department, Box 7625, Raleigh, NC 27695-7625, USA, E-mail: Sherman@unity.ncsu.edu, http://www.bae.ncsu.edu/people/ faculty/sherman.

Oasis Worm Farms, Norma Minto, P.O. box 15, West Union, IL 62477, USA, E-mail: oasiswormfarm@tigerpaw.com.

Oerther Family Farm, SE Sunnyside Road & 162nd, Clackamas, OR 97015-9351, USA.

Ogopogo Worm Farm, 2610 Glenmore Road, Kelowna, British Columbia, Canada- V1V 1B6, E-mail: ogopogowormfarm@shaw.ca.

Ohio State University Soil Ecology Department, Dr. Clive Edwards, 400 Aronoff Laboratory, 318 West 12th Avenue, Columbus, OH 43210, USA, E-mail: soilecol@osu.edu.

Olympic Worm Castings, Duane Van Zandt, 897 Mox-Chehalis Road, McCleary, WA 98557, USA.

Oregon Soil Corporation, Dan Holcomb, 1324 Beaver Lane, Oregon City, OR 97045, USA, E-mail: oresoil@aol.com.

Organic Renewal Enterprises, C/o Lower East Side Ecology center, P.O. Box 20488, New York, NY 10009, USA, E-mail: oreino@earthlink.net.

Organica Biotech Private Limited, 36, Ujagar Industrial Estate, Waman Tukaram Patil Marg, Govandi, Next to Pepsi Company, Deonar, Mumbai-400001, Maharashtra, India.

Original Vermitech Systems, ltd., Al Eggen, 2328 Queen Street East, Toronto, Ontario M4E 1G9, Canada, E-mail: aleggen@vermitechsystems.com, http://www.vermitechsystems.com.

Oscar's Bait Farm, Oscar & Irene Fincher, Rt. 3, Box 49, Stephenville, TX 76401, USA.

Ozark Worm Farm, Marvin Dickman, Rt. 3, Box 160, Willow Springs, MO 65793, USA.

Pacific Garden Company, Scott Subler, HCRI Box 150, Millheim, PA 16854, USA, E-mail: scott.subler@pacific-garden.com.

Pacific Southwest Farms, Barry Meijer, 13182 Baker Avenue, Ontario, CA 91761, USA, E-mail: Bmeijer@aol.com.

Parry Agro Industries Ltd. Tiam House Annexe, 2, Jehangir Street, Chennai-600001, Tamil Nadu, India.

Pat's Worm Ranch, Pat & Sonny Kellum, P.O. Box 3194, San Antonio, TX 78211, USA.

Patil Vermi Farm Behind Prasad Lodge, Gadag-582101, Karnataka, India.

Peaceful Valley Farm Supply, P.O. Box 2209, Grass Valley, CA 95945, USA, http://www.groworganic.com.

Pennie's Lane to Vermicomposting, Pennie Bernas, 325 Governor's Road, Paris, Ontario, Canada, E-mail: brandy@worldchat.com.

Pest Control Ltd., P.O. 3228, 479,5th Cross, HMT Layout, R.T. Nagar, Bangalore-560032, Karnataka, India.

Pinal Mountain Worm Farm, Don White & Dave Boerst, Westridge Drive, Rt. #1 Box 2F, Globe, AZ 85501, USA, E-mail: dblanco@theriver.com.

Pleasant Shade Worm Ranch, Jack McCathran, Rt. 1, Box 47, Pleasant Shade, TN 37145, USA.

Powell's Bait Farm, John W. Powell, Star Rt., Box 18, Wiseman, AR 72587, USA.

Purple Mountain Farms, John Fuller, 34750 Kooden Road, Winchester, CA 92596, USA,E-mail: purplemountainfarms@yahoo.com.

R P G Life Sciences Ltd. 4th Mile Stone, Sirsa Road, Hisar- 125001, Haryana, India.

Rabbit Barn, Larry Sigafoos, 630W, Clausen, Turlock, CA 95380, USA, E-mail: kxva66a@prodigy.com.

Rabbit Hill Farm, Jay & Joanne Mertz, 288 SW CR0020, Corsicana, TX 75110, USA, E-mail: rhf@airmail.net.

Raghuvar Agrotech Industries Sanjeevani Farms, Beawar Road, Ajmer-305001, Rajasthan, India.

Ramesh Bhai Organic Farm Kukama, Bhuj-370001, Gujarat, India.

Rancho Verma Farms, Matthew & Holiday Cordua, 26690 Greentree Avenue, Medera, CA 93638, USA, E-mail: mattnholiday@earthlink.net, http://www.ranchoverma.com.

Rashtriya Organic Fertilizers D/97, ambabari, Jaipur- 302012, Rajasthan, India.

Red Branch Farms, Deborah Bowers or Denise Grapes, 140 East Stetson Avenue #104, Hemet, CA 92543, USA, http://www.alcasoft.com/worms/.

Regional Bait & Marketing, Gary & Don Wolcott, 8006 Plank Road, Thomason, OH 44086, USA.

Resource Conversion Corp., Joseph Roberts, 7825 Fay Avenue, Suite 380, La Jolla, CA 92037, USA.

Rising Mist Organic Farm, Pure Mind Foundation, 1228 Westloop PMB 320, Manhattan, KS 66502, USA, E-mail: puremind@kscable.com, http://www. puremind.net.

Robert & Cindy Ambrose, 534 Pumphouse Road, New Florence, PA 15944, USA, E-mail: jed@connect-me.net.

Roberta (Bobby) Trombly, 3030 Marshall, Cincinnati, OH 45220, USA.

Room To Sqworm, Tracy Myhalyk, Boone, NC, USA, E-mail: Tracy@RoomToSqworm.com, http://www.RoomToSqworm.com.

Saveer Biotech Limited, D-54, Site-IV, Industrial Area, Surajpur, UPSIDC, Greater Noida-201306, Kasna, U.P. Phone No: +91 – 0120-4052500, Fax: +91 - 0120 – 4052510 Email: business@saveer.com.

Sha Agro Services 4833/24, Prahlad Lane, Darya Ganj, New Delhi-110002, India.

Shallow Creek Ranch, Steve Smith, P.O. Box 299, Foresthill, CA 95631, USA, E-mail: SCS@Foothill.net.

Sharkey's Worms, Lori Sharkey, 530 S.E. Baker Avenue, Gresham, WI 97080, USA.

Sharon's Worm World, Sharon McLachlan, Ramon, CA, USA, E-mail: sharonswrm@aol.com.

Shield's Publications, Patrick Shields, P.O. Box 669-W, Eagle River, WI 54521, USA, E-mail: shields@nnex.net, http://www.wormbooks.com.

Shri Chaudhari Associates H-21, Madhuvan Colony, Kisan Marg, Tonk Phatak, Jaipur-302012, Rajasthan, India.

Shri Ganganagar District Milk Union, Milk Union Processing Plant, URMUL Bhavan, PB No. 55, Sriganganagar Road, Bikaner-334001, Rajasthan, India.

Shri Gaushala, Pilani, Jhunjhunu-333001, Rajasthan, India.

Shri Kamdhenu Vaidik Gaushala, Pransala, Rajkot-360001, Gujarat, India.

Shri Krishan Gaushala, Chirawa, Jhunjhunu-333001, Rajasthan, India.

Shri Mahaveer Gaushala, Malkisar Station, Tehsil-Loonkaransar, Distt. Bikaner-334604, Rajasthan, India.

Shri Ram Salvent Extractions Pvt. Ltd. Kashipur Road, Jaspur-496338, Uttaranchal, India.

Shri Ram Solvent Extactions Pvt. Ltd., Kashipur Road, Jaspur, Udam Singh Nagar-244712, Uttaranchal, India.

Sisodiya Manure Milses, 26-27, Trade House, 3rd Floor, Ashawani Bazar, Udaipur-313001, Rajasthan, India.

Siyaram Manure Siyaram Street Durgapura, Jaipur-496338, Rajasthan, India.

Solana Recyclers, 137 N. El Camino Real, Encinitas, CA 92024, USA, E-mail: Solana@adnc.com, http://www.digitalseed.com/solana.

Solano Worm Farm, 18570 NW 43rd Ct. Road, Citra, FL 32113, USA.

Soman Bhai Garsisa Farm Mandavi, Bhuj-370001, Gujarat, India.

Sonoma Valley Worm Farms, Jack & Lois Chambers, Sonoma, CA, USA.

Southern Worm Enterprise, Susan Damper, 12118 Marilyn Lane, Hammond, LA 70403, USA.

Speegleville Worm Farm, Bob & Kathy Pettigrew, 303 Randy Drive, Waco, TX 76712, USA, E-mail: red_wigglers@yahoo.com.

Spiker Work and Casting, Don Spiker, 12877 North 600th Street, Wheeler, IL 62479, USA, E-mail: dspicker@mmtcnet.com, http:// www. spikerwormandcasting.com.

Sri Gayathri Biotec, D. no 25-10-26, Telagapalem, Ponnur-522124, Andhra Pradesh, India.

Srushti Bio-organics B/3, Hirappana Apt. Nr. Vijay Restaurant, Navrangpura, Ahmedabad-380009 Gujarat, India.

Stages Worm Farm, Larry Staggs, 456 Yermo Road, Yermo, CA 92398, USA.

State Bio-Logical Pest Control Lab, Department of Agriculture, ATMA Project, Shimla-171001, Himachal Pradesh, India.

State Department of Agriculture, Govt. of Uttar Pradesh, Information Bureau.

State Directorate of Agriculture, Indore Division, Indore and Agriculture College, Indore-452001, India.

Sterling Farm Research & Service Pvt. Ltd. P. Box. No. 2344, Kochi- 682016, Kerala, India.

Stewart's Worm Farm, Marie Stewart, P.O. Box 188, Springhill, LA 71075, USA.

Stollar (Indian) Pvt. Ltd. 407, Sureja Tower-I, District Centre, Janakpuri, New Delhi-58, India.

Sukhakarta Agrotech, Rane nagar, Nashik-422007, Maharashtra, India.

Suman Vermicompost, Sadabad Road, Jalesar, District Etah, Uttar Pradesh - 207302, India.

Supa Biotech Pct.Ltd. Kohinoor Building Dina Lodge Compound, Mallital, Nainital-263139, Uttarakhand, India.

Superior Bait Farm, Michael Stockman, 7701 Silver Lake Road, Wolverine, MI 49709, USA.

Swati Enterprises, Mrs. Swati Sharma (Proprietor), No. 144 A, Pocket A, Dilshad Garden, New Delhi - 110095, Delhi, India.

T & I Global Ltd.19-RN, Mukherji Road, Kolkata-700001, West Bengal, India.

Tate's Worm & Fern Farm, Robert E. Tate, Rt. 1, Box 503, Chatom, AL 36518, USA.

TERI (The Energy Research Institute), Darbari Seth Block, Habitat Place, Lodhi Road, New Delhi-110003, India.

Texas Organics Pvt. Ltd. M.C. Road, Pandalam, Pathanamthitta-69523, Kerala, India.

Thanvi Farm Phaloudi, Jodhpur-342001, Rajasthan, India.

The Chef's Garden, 973 County Road 8, Heflin, AL36264, USA, http://www.chefsgarden.com.

The Cultured Worm, Lee Barber, Durham, NC, USA, E-mail: lbarber@ duke.edu.

The French Institute for Scientific Research for Cooperative Development (ORSTOM):Dr. Patrick Lavelle, Centre de Bondy, 72, Route d'Aulnay, 95143 bondy Cedex, France.

The Garbage Man, Inc., Steve Logue, P.O. box 1752, Columbia, TN 38402, USA.

The Happy-O Worm Farm, 11719 Hesby Street, Valley Village, CA 91607, USA, E-mail: hapyo@worldnet.att.net.

The McNelly Group, Jim McNelly, P.O. Box 7444, St. Cloud, MN 56302, USA, E-mail: naturtech@composter.com.

The National Agricultural Library (NAL), 10301 Baltimore Blvd., Beltsville, MD 20705, USA.

The Squirmy Wormz Farm, Linda Holland, Australia, E-mail: Linda@suirmy-worms.com, http://www.squirmy-worms.com.

The Wigglin N., T. Noland, Beggs, OK, USA, E-mail:Tnoland@worldnet, http://home.att.net/~tnoland/index.htm/.

The Worm Box, 315 Lewis Street, Moscow, ID 83843, USA.

The Worm Concern, Bill Harding, 4897 Wanda Lane, Forest Ranch, CA 95942, USA, E-mail: freedude40@aol.com.

The Worm Factory, 129D Sherbrooke Street, East, Perth, Ontario, Canada, E-mail: wormfact@perth.igs, http://www.thewormfactory.net.

The Worm Lady, Marilyn Charest, 32912 Buttercup Lane, Squaw Valley, CA 93675-9774, USA, E-mail: wormlady@spiralcom.net.

Three Trees Farm, Chris. Boissevain, 73470 Abeene Lane, Cottage Grove, Or 97424, USA.

Tip Top Bio-Control, P.O. Box 7614, Westlake Village, CA 91359, USA, E-mail: sales@tiptopbio.com.

Topline Wholesale Distributing Co., Rea Blossom, P.O. Box 45535, Rio Rancho; NM 87174-5535, USA, E-mail: topline@topline-2000.com, http://www.topline2000.com.

Trees for Life-Green World Associates 620, Sector-19, Pocket-II, Dwarka, New Delhi 110045, India.

TriFam Earthworm Farm, 8749 East Wilson Road, Bakersfield, CA 93307, USA, E-mail: trifam@bigfoot.com.

Triformis Corp., Los Angeles, USA, E-mail: info@triformis.com.

Trinity Ranch, Bob Ingram, 5750 Duda Rd., House springs, MO 63051, USA, E-mail: trinitywigglers@peoplepc.com, http://www.trinity-ranch.com.

Trio-Chem Techno-legal Services Pvt. Ltd. "Ishwari", 17-A, Gokul Society, Navi Peth, Pune-411030, Maharashtra, India.

Triple Z Ranch, 1639 Calzada Avenue, Santa Ynez, CA 93460, USA.

TTDC – Project Management Unit (PMU) FRDC Branch, Uttaranchal Secretariat, Dehradun-248001, Uttarakhand, India.

Udaipur District Milk Union, Milk Union Processing Plant, Udaipur-313032, Rajasthan, India.

Uncle Jim's Worm Farm, Jim Shaw, RD#1, Box#1537, Lehman Road, Spring Grove, PA17362, USA,E-mail: jshaw59788@aol.com, http://www.unclejim.com.

Unco Industries, Tom Chapman, 7802 Old spring Street, Racine, WI 53406, USA.

University of Agricultural Sciences Dharwad-580001, Karnataka, India.

University of Stellenbosch, Dr. Adrian Reinecke, Department of Zoology, Private Bag X5018, 7599 Stellenbosch, South Africa, E-mail: cb4@maties.sun.ac.za.

UP AGRO Industries Corporation, Hapur-245101, Uttar Pradesh, India.

UP Diversified Agriculture Support Project Block-B1B2, IV Floor, PIC UP Bhawan, Gomati Nagar, Lucknow-226001, UP India.

Uttar Pradesh Land Development Corporation TC/19 B, Vibhuti, Gomati Nagar, Lucknow-226001, UP India.

Valerie Cowperthwaite, 17 Poronui Street, Mt. Eden, Auckland, New Zealand, E-mail: valereie@stuart.ak.planet.co.nz.

Valley Worm Growers, John Banks or Richard Hicks, 100A San Bernardino Road, Ridgecrest, CA 93555, USA, E-mail: mercury@ridgecrest.ca.us.

VanArsdale's Worm Farm & Sales, Frank VanArsdale, San Jacinto, CA 92582, USA, E-mail:wormguy@ivic.net, http://www.ivic.net/~wormguy.

Vanashree AgricultureJ-116, Megacentre, Magarpatta, Hadapsar, Pune-422001, Maharashtra, India.

VANRAI, 2064, Vijaynagar, Pune-411030, Maharashtra, India.

Vatika Farm, Vatika Road, Sanganer, Jaipur-496338, Rajasthan, India.

Vermi-Cast, Walter Larsen, 2052 Hidden Creek Street, Corona, CA 92881, USA, E-mail: larsenw@hninc.com, http://www.vermi-cast.com.

Vermico, Peter Bogdanov, P.O. Box 2334, Grants Pass, OR 97528, USA, E-mail: vermico@vermico.com, http://www.vermico.com/.

Vermicoast, Shelley Grossman, 1387 Basswood Avenue, Carlsbad, CA 92008-1904, USA, E-mail: shelley@adnc.com, http://www.vermicoast.com.

Vermicomposting Systems, David Lee Hoffman, P.O. Box 10, Lagunitas, CA 94938, USA.

Vermiculture Services International, Richard Morhar, 581 Camino Manzanas, Thousand Oaks, CA 91360, USA.

Vermitechnology Unlimited, Larry Martin, P.O. Box 130, Orange Lake, FL 32681, USA, E-mail: vermin@aol.com, http://www.vermitechnology.com.

Vigyan Samti, Road No. 17, Ashok Nagar, Udaipur-313001, Rajasthan, India.

Vijaylakshmi Insecticides and Pesticides Ltd., Ethakota, Andhra Pradesh-533238, India.

Viscor Distribution, Inc., 4291 Meridian Street, Suite 402, Bellingham, WA 98226, USA, E-mail: canowormsinfo@viscor-dist.com, http://www.viscor-dist.com.

Vishva Bharti Kisan Jagriti Shikshan Sansthan, Shri Ladva Gaushala, Ladva, Hisar-125087 Haryana, India.

Vishwa Bharati Kisan Jagriti Shikshan Sansthan, Shree Ladwa Goshala, Ladwa, Hisar-125087, Haryana, India.

Visva Bharati Kisan Jagriti Shiksha Sansthan, Shree Ladwa Goshala, Ladwa Hisar-125087, Haryana, India.

Wally's Worm Ranch, Mary Lee Allen, 101-B, West McKnight Way #245, Grass Valley, CA 95949, USA, E-mail: mlallen@jps.net.

Watershed Association- Sawar, Saiwar, Jamwaramgarh Road, Jaipur-302001.

Way Out Worms, Victoria Graham, 10203 Barnett Valley Road, Sebastopol, CA 95472, USA, E-mail: WOWFarm@webtv.net.

Wiggly Worm Wranglers, Grayson, KY 41143, USA, Email: wiggly wormwranglers@hotmail.com.

Williamson Worm Farm, Bobby & Lisa Williamson, 1661 Brydville Freeman Road, Delco, NC 28436, USA, E-mail: rrwilli@bwwonline.com.

Willies Organic Recycle Machine Systems, Den and Diane Phillips, 1426 East 7th Street Suite A, National City, CA 91950, USA, E-mail: densworms@freeyellow.com, http://www.freeyellow.com/members/ densworms/.

Willingham Worm Farm, Sybil Willingham, Rt. #1, Box 241, Butler, GA 31006, USA.

Wilson's Worms, Joe Wilson, P.O. Box 331, Lincoln, MT 59639, USA, E-mail: jwilson@linctel.net, http://www.wilsonsworms.com.

Wonder Worms, John & Bonnie Mathis, 1014 Kyle Court, Ridgecrest, CA 93555, USA, E-mail: jbmathis@ridgecrest.ca.us.

Woody Thomas and Eric Hughes, 6111 Claypoint Road, Chesterfield, VA 23832, USA.

Worm Digest, S. Zorba Frankel, P.O. Box 544, Eugene, OR 97440-0544, USA, E-mail: mail@wormdigest.org, http://www.wormdigest.org.

Worm Discount Store, Anne Menefee, 7500 Amber Road, Fort Wayne, IN 46814, USA, E-mail: Mene111@aol.com.

Worm World, Inc, Harry Windle, 12425 NW C.R. 231, Gainesville, FL 32609, USA, E-mail: windle@gnv.fdt.net, http://gnv.fdt.net/~windle/.

Worm World, Kay & John Inhat, 26 Inhat Lane, Avella, PA 15312-2106, USA, E-mail: wormw@access.hky.com, http://www.wormwrld.com.

Worm's Way, 7850 N. Highway 37, Bloomington, IN 47404, USA, http://www.wormsway.com.

Worms A-Lure, Connie & John Foht, 194 Ansel Road, Geneva, OH 44041, USA.

Worms by Weber, Gleen & Linda Foster, P.O. Box 1126, Yellville, AR 72687, USA.

Worms, Supplies and Information, North Carolina Cooperative Extension Service, Rhonda L. Sherman, Extension Solid Waste Specialist, College of Agriculture and Life Sciences, Department of biological and Agriculture Engineering, Box 7625, Raleigh, NC 27695-7625, USA.

Worms' Work Farm, Tim Berlin, 5350 NC Hwy 194 North, Boone, NC 28607-1461, USA,E-mail: wormswork@yahoo.com.

Worms-R Us, Bill Williams, Aguanga, CA, USA.

Wright Worm Farm, Don & Bobbie Wright, 32205 Meadow Blossom, Nuevo, CA 92567, USA, E-mail: wriwormf@gte.net, http://www. wrightworm-farm.com/.

Yahoo Compost, Fernando Pastor, San Francisco, CA, USA, E-mail: quesapastor@yahoo.com.

Yahoo Herb an' Ecology, 968 Valencia Street, San Francisco, CA 94110, USA.

Yashovati Earthworms and Pest Research Institute, Tulsiram Road, Tinsukiya-786125, Assam, India.

Yashovati Earthworms and Pest Research Institute, Tulsiram Road, Tinsuliya-786125, Assam, India.

Yelm Earthworm & Castings Farm, 14741 Lawrence Lake Road SE, Yelm, WA 98597, USA, E-mail: mail@yelmworms.com, http://www. yelmworms.com.

ANNEX III: IMPORTANT WEBSITES RELATED TO VERMITECHNOLOGY

1. www.allorganiclinks.com
2. www.indiamart.com
3. www.organicpathways.co.nr
4. www.newscientist.com
5. www.intracen.org/mds/sectors/organics
6. www.webfoodpros.com
7. www.organicsnewzealand.org
8. www.fasusdagov/info/agexporter
9. www.agr.ca
10. www.foodcontach.com
11. www.nre.vicgov.av
12. www.ocia.org
13. www.ecomall.com
14. www.linksorganic.com
15. www.gepa.org.ge
16. http://www.vanashreeagrotech.com
17. http://www.niir.org
18. http://www.ativabio.com
19. http://www.srigayathribiotec.com

INDEX

D

E

F

G

N

O

P

Q

R

S

V

W

Y

Z